ALL SHALL HIDE

by Taylor A. Cisco, Jr.

AuthorHouse™
1663 Liberty Drive
Bloomington, IN 47403
www.authorhouse.com
Phone: 1-800-839-8640

© 2010 Taylor A. Cisco, Jr. All rights reserved.

No part of this book may be reproduced, stored in a retrieval system, or transmitted by any means without the written permission of the author.

First published by AuthorHouse 9/13/2010

ISBN: 978-1-4520-5636-4 (e)
ISBN: 978-1-4520-5634-0 (sc)
ISBN: 978-1-4520-5635-7 (hc)

Library of Congress Control Number: 2010910602

Printed in the United States of America

Unless otherwise indicated, the Scriptures quoted are taken from the King James Version of the Holy Bible.

Because of the dynamic nature of the Internet, any Web addresses or links contained in this book may have changed since publication and may no longer be valid. The views expressed in this work are solely those of the author and do not necessarily reflect the views of the publisher, and the publisher hereby disclaims any responsibility for them.

This book is dedicated to my wife, Yvonne Lorraine Hipps-Cisco. She has been my sole soul-mate since 1972. Yvonne persistently encouraged me to compile this book from my bible prophecy presentations.

Table of Contents

Figure Credits ix

Preface xi

Introduction xv

Chapter 1 – Sixth Seal Earthquake 1
    Section 1-1. Heavenly Command Center 1
    Section 1-2. Warning Signs for Earthquakes 2
    Section 1-3. Definitions 2
    Section 1-4. Great Earthquakes 3
    Section 1-5. Implications 5

Chapter 2 – The Sun Becomes Black as Sackcloth of Hair 7
    Section 2-1. Historical Trustworthiness of the New Testament 9
    Section 2-2. Crucifixion Darkness Accounts 10
    Section 2-3. Explanatory Attempts 15
    Section 2-4. Acute Solar Darkening Events Hypothesis 18
    Section 2-5. Test of Hypothesis 23
    Section 2-6. Deductions 31
    Section 2-7. Black as Sackcloth of Hair 35

Chapter 3 – The Moon Becomes As Blood 37
    Section 3-1. The Black Sun With Red Moon Sign 37
    Section 3-2. Lunar Luminescence 38

Section 3-3. Wide Area Reddening..46
Section 3-4. Analyses of Lunar Rocks 52

Chapter 4 – Stars of Heaven Fall to Earth 55
Section 4-1. Identifying the Falling Stars.......................................55
Section 4-2. Wingless Aircraft 58
Section 4-3. Wingless Aircraft Propulsion Systems 60
Section 4-4. What Causes the Points of Light to Fall 63

Chapter 5 – Heaven Rolls Back..67
Section 5-1. The Great Auroral Storm of 1859 68
Section 5-2. The Carrington Event 69
Section 5-3. Ice Cores..75

Chapter 6 – Mountains Are Moved...81
Section 6-1. Physical Implications of Shifted Mountains 81
Section 6-2. Heliophysically Triggered Dislodgements................83
Section 6-3. Biblical Indications of Water Beneath Mountains ...83
Section 6-4. Evidence of Water Beneath Mountains and Islands 89

Chapter 7 – The Great Migrations 93
Section 7-1. All Will be Able to Hide..93
Section 7-2. Creating Enough Shelters before the Sixth Seal Events 98
Section 7-3. Penetrating Earth's Shields.....................................102
Section 7-4. Aerospace Hazards...103
Section 7-5. Ground Level Hazards..105

Chapter 8 – Correct Interpretation of Events 117

Chapter 9 – The Moral...123

Index...127

Endnotes..133

Figure Credits

Cover Figure – This illustrates the relative size of the Earth compared to the Sun and the large coronal mass ejection it had launched during October 2003. The image was captured by the international SOHO spacecraft and is credited to both the National Aeronautics and Space Administration (NASA) and the European Space Agency (ESA).

Figure 1, Page 20 – This image by NASA displays the granulation of the convection cells in the Sun's surface (the photosphere) and the darker regions are known as sunspots. The umbra is the darker region of a sunspot and the penumbra is the lighter area surrounding it. Light bridges are the narrow lines of granules passing through the umbra. Images like this one have been used to illustrate Tilt-Joy's Law: leading spots are closer to the solar equator than trailing spots in sunspot groups.

Figure 2, Page 74 – The Carrington Flare. Carrington, R. C. (1859, November). Description of a singular appearance seen in the Sun on September 1, 1859. *Monthly Notices of the Royal Astronomical Society*, **20**, 13. Provided by the NASA Astrophysics Data System.

Figure 3, Page 78 – This National Science Foundation photo was taken by Jonathan Berry on May 10, 2002, of the Aurora

Australis, or Southern Lights, over the NOAA Atmospheric Research Observatory (ARO) at the Amundsen-Scott South Pole Station. The line of flags marks the path to the ARO.

Figure 4, Page 79 – NASA photograph of the Aurora Australis above the Earth taken by an astronaut aboard Space Shuttle (STS-39) in 1991. Estimated altitudes for the aurora lights ranged from 50 to 80 miles.

Figure 5, Page 101 – U. S. Air Force tunnel boring machine at Little Skull Mountain, Nevada. Source: U. S. Department of Energy, Nevada Operations Office; December 14, 1982.

Figure 6, Page 113 – Group sheltering in a large building (Fig. 10.23). U. S. Department of Defense and the Office of Civil Defense (1968). *Radiological Defense Text Book* [p. 139]. Washington, D.C: United States Government Printing Office.

Figure 7, Page 114 – Family prefabricated and plywood backyard shelters and basement corner shelter (Fig. 10.27b,c,d). U. S. Department of Defense and the Office of Civil Defense (1968). *Radiological Defense Text Book* [p. 140]. Washington, D. C: United States Government Printing Office.

Figure 8, Page 115 – Deep Shaft Structure. Kao, A. (1985, April). Literature survey of underground construction methods for application to hardened facilities [p. 34]. Technical Report M-85/11 U.S. Army Corps of Engineers. (AD Number ADA 155212)

Preface

All Shall Hide was the product of several compulsions. These served as the driving forces that had enabled me to persevere to complete this work.

First and foremost, I did not want God to judge me guilty of not sharing with others the variety of phenomena I had discovered. The facts I had learned promoted the physical interpretations of biblical prophecy. Failure to inform others about these findings would have placed me in danger of violating one of God's principles regarding obligations to warn people (Ezekiel 33:1-9). The discoveries that have been compiled can facilitate understanding and strengthen belief in the *Holy Bible*.

Secondly, this book was written in partial compliance with my wife's pleas and the suggestions from my children and friends to compile a book based on my bible prophecy presentations. I had prepared handouts and a booklet for various prophecy workshops. The booklet addressed twenty-six prophecies. My wife and friends wanted me to expand it from an outline format into a series of explanations. This book addresses only one of those twenty-six topics – Revelation 6:12-17.

Thirdly, it was produced to serve as a resource for helping teachers and film makers provide a more accurate depiction of biblical descriptions of the future. I have been annoyed with portrayals of the darkness at the crucifixion as a solar eclipse or of the trumpet and bowel of wrath events occurring before the Sixth Seal events of Revelation. And, I have been amazed with the misunderstanding of the sequence of events leading up to the Day Wrath. To clarify such sequences was a major goal.

The fundamental concept of this book was realized during the 1980s while I was answering questions from twins about the signs for the Great Tribulation. The popular answer back then, in fundamentalist churches, was the abomination of desolation performed by the

Antichrist in the restored temple of Jerusalem. It occurred to me as I mentally structured the answer for their young ears that there was a larger and more significant sign that had received little attention. It had been introduced by Joel (Joel 2:28-32), recited by Peter (Acts 2:16-21), and explained further by John (Revelation 6:12-17). That sign was the blackening of the Sun and the concurrent reddening of the Moon. It has been presented in the *Holy Bible* as the great precursor of what has been called the Day of Wrath, the Wrath of the Lamb, and the Day of the Lord. This was a sign that would attract more attention than the abomination of desolation. That solar darkening event shall be more severe than storms of global sunspots; will be directly observed, without the aid of news media, by everybody on or near the Earth; and, after a great earthquake, will cause all of the populations of the world to flee to cliffs, caves, and underground shelters. The literal solar blackout did not seem preposterous to me because I had remembered reading about a starspot that had covered over half of the surface of one of the components in the Lambda Andromedae binary star system. If it could happen out there, it could happen here.

I had studied explanations which had attributed that future solar blackout to clouds of smoke and debris generated by volcanic eruptions, meteor impacts, and nuclear wars. But, during that moment in time it took to select the right words the youngsters could understand, I realized the knowledge I possessed could be used to explain a literal darkening of the visible portion of sunlight. Their questions placed me on a path of research that yielded workshops, bible class lectures, a dissertation, Wikipedia articles, my first sermon, and this book about the physically most dramatic set of phenomena that shall herald the beginning of the Day of Wrath.

The decision to commit to writing this book was based on a form of fleece I submitted to the Lord during the Spring of 2002. Through prayer, I indicated to Him I would write a book about the six seal events if I could find some scientific evidence supporting the phenomena of the entire Moon becoming red under a blacken Sun. I had presumed some sort of irradiance from the bombardment of the lunar surface by atomic particles from the Sun may be the cause. But, I did not want a false presumption to place me in a wild goose chase. I went to the books I had collected while a teenager. During the sixties and seventies, I had read only the chapters and sections in many books that had pertained to rockets. The first book I examined was Willy Ley's *Ranger to the Moon*. I was astounded to discover it had a chapter entitled "The Mystery of the Red Spots." I read it. It was about the reports of glowing red spots on the Moon observed by Sir William Herschel in 1783 and James Greenacre and Edward Barr in 1963. I checked a second source. I had been an active member of the Twin Cities Amateur Astronomers club during the sixties and its club magazine had been *Sky & Telescope*. I dusted off my stack of old issues and found the December 1963 article by Greenacre and then stumbled into March 1964 article by Zdeněk Kopal and Thomas Rackham on the red colored, wide area luminescence of the Moon. That was it – my fleece was sopping wet. I made the commitment to write the book and began to search for material that could be incorporated with proofs of concept justifying literally physical interpretations of Revelation 6:12-17.

The final product became a compilation of reliable historical records, analyses from solar physics, and recent findings from clinical cosmobiology. These became the foundations for the principle concepts. Hypothesized intensification of those concepts were then used to explain the phenomena caused by the blacken Sun. Many statements were referenced by endnotes because the majority of principle concepts were controversial (e.g., lunar luminescence and clinical cosmobiology). Subsequently, over ninety percent of the sources cited were either peer reviewed journal articles or books by reputable publishers. And, the self-publishing works cited had been written by credentialed authors.

Chicago area libraries, book stores, and the Internet were the source of the literature used in this study. Materials were obtained through the Enter-library Loan Department, Oak Park Public Library; the John Crerar Library, University of Chicago; the Harold Washington Chicago Public Library; and the Triton College Library, Melrose Park. Many books and papers were extracted from the web sites of the National Aeronautic and Space Administration (NASA) Astrophysics Data System, electronic preprint archives of arXiv.org, American Institute of Physics (AIP) Conference Proceedings, Defense Technical Information System (DTIS), Googlebooks.com, and the National Technical Information Service (NTIS). Eclipse predictions (general locations, dates and duration) for total solar eclipses were by Fred Espenak and Jean Meeus, NASA Goddard Space Flight Center, Greenbelt,

Maryland, and were obtained from the "Five Millennium Catalog of Solar Eclipses" at the NASA Eclipse Home Page.

Numerical indices of *The New Strong's Exhaustive Concordance* were used to denote the Greek roots of words examined in this study. Definitions and usage for particular Greek roots were taken from Blue Letter Bible web site. The root words were rarely expressed in terms of the Greek alphabet. For example, in Revelation 6:15, the Greek root for rocks is represented by [SEC 4073] and is used to mean cliffs.

First and foremost, I must express my gratitude to God (i.e., Yahweh, Jesus Christ, the Holy Ghost) for His sequence of events that promoted the creation of this document. Secondly, I must honor my wife, Yvonne Cisco, for her insistence that I convert my Bible prophecy presentation notes into a book. And, for the time and effort she and our daughter, Catherine Cisco, and son, Taylor Cisco, III, spent retrieving journals, purchasing books, and discussing critical aspects of the book. Additional insight and discussions were provided by Dr. Charles A. Perry, U. S. Geological Survey, Lawrence, Kansas; Dr. Donald V. Reames, NASA Goddard Space Flight Center, Dr. Robert Reedy, Los Alamos National Laboratories; Dr. Shenghui Li, Department of Earth and Space Sciences, University of Washington, Seattle; David B. Williams, AAVSO President, James Fuller, Sales Representative, AuthorHouse Publishers. Expressions of gratitude must be given to Grace Lewis, Enter-library Loan Department, Oak Park Public Library, Illinois; and Barbara Pope, Request Coordination Office Manager, NSSDC and WDC-A/R&S, NASA/Goddard Space Flight Center, Greenbelt, Maryland. Spiritual encouragement, guidance, and support were provided by Dr. Gloria J. Forward, President, International Apostolic University of Grace and Truth, Indianapolis, IN; Bishop Earl O. Holiman, Bethsaida Temple Christian Center, Denver, Colorado; Bishop Charles E. Davis, Diocesan of the Sixth Episcopal District of the Pentecostal Assemblies of the World; Dr. Joyce Walker, Vice President, and Dr. Margaret Wright, Dean, God's Bible Institute, and Lawrence Kearney, Sunday School Superintendent, Indiana Avenue Pentecostal Church, Chicago, Illinois. Special thanks must be extended to Minister Derrick Thurston, Apostolic Faith Church. And, tremendous gratitude goes to my mother, Julia Cisco, and my late father, Taylor Cisco, Sr., for providing a home that nurtured biblical eschatological research.

Introduction

If you enjoy reading about literal interpretations of biblical prophecy, you will enjoy and find this book to be a very useful resource. *All Shall Hide* provides a fresh examination of the awesome and terrifying events that precede the Day of Wrath as described in Revelation 6:12-17.

Very few eschatological works have identified the forces that frighten all of the populations of the world to flee to caves, mountains, cliffs, and underground facilities after a great earthquake. People who have survived earthquakes normally sleep outside, regardless of weather conditions. What in the future shall drive them underground? *All Shall Hide* answers that question in terms of summaries from collaborative, clinical studies. They have identified correlations between incidences of sudden cardiac death and myocardial infarctions with intensified geomagnetic and cosmic ray (neutron flux) activities. These stem from the corollary that a darken Sun is not a quiet Sun.

All Shall Hide summarizes the trustworthy historical records of the Sun blackening. Their descriptions cannot effectively be attributable to eclipses, cloudy weather, or volcanic debris. For example, the biblical and non-biblical accounts of the darkness at the crucifixion were examined. The descriptions of the solar blackout recorded by Herodotus, Plutarch, and Aristides that occurred in BC 480, a few months before the famous Battle of Thermopylae, were reviewed. These and other reputable historical records were used in the context of Solomon's saying; there is nothing new under the sun (Ecclesiastes 1:9) to argue for the reality of future, solar blackouts.

Another question, why are meteor showers an inappropriate explanation for the scripture: "And the stars of heaven fell unto the earth, even as a fig tree casteth her untimely figs, when she is shaken of a mighty wind" (Revelation 6:13)? The answer: the extended flight paths of falling figs do not diverge from radial points. Meteor showers are named after the constellations that contain their radial points. *All Shall Hide* provides an explanation that is consistent with the biblical details.

All Shall Hide is consistent with the following: "All scripture is given by inspiration of God, and is profitable for doctrine, for reproof, for correction, for instruction in righteousness: That the man of God may be perfect, throughly furnished unto all good works." (2 Timothy 3:16-17). Prophecy, like the other classifications of scripture, should be studied.

Chapter 1
Sixth Seal Earthquake

> And I beheld when he had opened the sixth seal, and, lo, there was a great earthquake; and the sun became black as sackcloth of hair, and the moon became as blood; (Revelation 6:12a,b)

Section 1-1. Heavenly Command Center

Verse twelve indicates the future events shall be caused by decisions made in Heaven. John recorded a series of experiences that resemble transactions in a military command center. For example, commanders stationed in the underground Launch Control Center of Davis-Monthan Air Force Base, Arizona, were able to flip switches and push buttons that could cause Titan II intercontinental ballistic missiles to deliver nuclear warheads to devastate cities over 5,000 miles away. In a similar manner, Jesus Christ will open the six seals on a scroll to cause dramatic physical events to transpire on the Earth, Sun, and Moon. Officers have used telecommunication systems to relay the images of forces following their orders. Jesus will use an infinitely sophisticated technology to convey the images of powers obeying His commands. The sights and sounds John recorded denoted the ambiance of impending battles and war.

It is interesting to note that theorists have mathematically proposed the existence of multi-universes.[1] Heaven could be one of those universes. We do not know its location, but we do know that John recorded a chain of events that would be started by the opening of the Sixth Seal in Heaven. God will determine when the fulfillment of this prophecy is to commence. Subsequently, this set of events will start with very little to no warning. It shall be an anomaly. Those of us in Heaven who are allowed to watch Jesus Christ open the Sixth Seal will know the great sign is about to rock the Earth. We, who have not died and are alive on Earth, will be surprised with the sudden emergence of the wide spread and strong earthquake.

Section 1-2. Warning Signs for Earthquakes

Researchers have been investigating various phenomena that may serve as triggers and/or precursors for earthquakes. Networks of monitoring stations, communications facilities, and computation agencies have been established to study and predict earthquakes. Well water levels, animal behavior, low frequency electromagnetic signals, tidal forces stemming from the relative positions of the Sun, Earth, and Moon, radon gas concentrations, land surface deformations, and the interplanetary magnetic field are some of the phenomena that have been monitored. Scientist have made progress towards identifying the type of changes in electromagnetic activity within and above the Earth that precede earthquakes.

The precursors for the unique and great Sixth Seal earthquake will emerge too rapidly for long range forecasting and early warning systems. The scriptures indicate the Day of the Lord shall take mankind by surprise: "For yourselves know perfectly that the day of the Lord so cometh as a thief in the night" (1 Thessalonians 5:2).

Section 1-3. Definitions

Revelation 6:12 contains the first description of an earthquake in the Book of Revelation. The word "earthquake" was translated from the Greek word *seismos*. It appears as a Greek root in words such as seismology, the science of earthquake phenomena, and helioseismology, the science of massive vibrations in the Sun. *Seismos* incorporates the Greek root *seio* [SEC 4578], meaning to shake, chiefly with the idea of concussion. And, it has been used in the *Holy Bible* to describe a great tempest at sea (Matthew 8:24).

Subsequently, the Sixth Seal earthquake does not have to be confined to the geologic definition – a shaking of the ground caused by abrupt shifts of rock along fractures in the Earth. This scripture may pertain to violent ground shocks that may be caused by a meteoroid impact, great explosions, volcanic eruption, a swarm of earthquakes, and/or, a great earthquake and it could pertain to storms and unusually strong winds. It should be noted that the fulfillment of this scripture shall be more unique than previous great earthquakes.

Section 1-4. Great Earthquakes

The following examples of these various forms of concussive phenomena are given to help manifest the unique features of the Sixth Seal earthquake:

A few earthquakes have disrupted large areas of land. On March 28, 1964, "Good Friday," the strongest earthquake to occur in North America struck 80 miles east of Anchorage. The energy from the magnitude 9.2 earthquake was felt over an area of 500,000 square miles. It snapped off the tops of trees near the epicenter and over 12,000 square miles of the earth's surface was lifted. Because of the small population density and the holiday, the death toll was limited to 115 in Alaska and 16 in Oregon and California.

The course of the Mississippi River was reversed near New Madrid, Missouri, by a series of three great earthquakes. The first one hit with a magnitude of 8.4 on December 16, 1811. The second one hit on January 23, 1812, with a magnitude of 8.6, and the third on February 7, 1812, with a magnitude of 8.8. Over 1,870 shocks were felt 200 miles away in Louisville, Kentucky from December 16, 1811 to March 16, 1812. Over 40,000 square miles of land was shaken. A 6,000 square-mile strip of land became known as the "Sunken Country." It was covered by water when that area sunk between 3 to 9 feet. Neither one of these have dislodged all of the mountains of Earth. But, the Earth did vibrate like a bell for three weeks after the quake.[2]

The earthquake that had great impact on Christendom occurred on Sunday morning, November 1, 1755, in Lisbon, Portugal. The first Magnitude 8 shock struck at 9:40 A.M. It lasted slightly more than six minutes and caused all public buildings and 12,000 dwellings to fall on and kill 30,000 people. The second and third great earthquakes struck at 10:00 A.M. and Noon. The subsequent tsunami and six days of fires raised the death toll to 70,000. Damage was reported as far away as 700 miles east of Lisbon. Ellen G. White quoted a report by Sir Charles Lyell about the largest mountains in Portugal being shaken and ruptured at their peaks to issue fire.[3] Many of the victims were killed in church. That catastrophe "affected religious and philosophical arguments in Europe for decades."[4] It retained a position of significant importance in the eschatology of Seventh Day Adventists.

The strongest earthquake in recent history struck Chile with a series of shocks from May 21 – 30, 1960. The strongest shock attained a magnitude 9.5 on the Richter scale. Over 5,700 in Chile, Hawaii, Japan, and the Philippines were killed by it. Volcanic eruptions, tsunamis, and the creation of islands, mountains, and lakes were generated by it.[5] A rupture in the Earth's crust that ran 600 miles along the coast of South America had caused it.

The deadliest earthquake in history occurred on January 23, 1556, in Shaanxi Province, China. Over 830,000 were killed in that Magnitude 8 earthquake. China was the victim of the deadliest earthquake of the Twentieth Century. Tianjin was struck on July 27, 1976, with a Richter scale 8.0 earthquake. Its estimated death toll reached 655,000. A large death toll is not one of the features of the Sixth Seal earthquake. The massive loss of life has been predicted in Revelation for other, future earthquakes. The Sixth Seal earthquake seems to serve more as a sign than as an act of judgment. Large death tolls are not one of the features explicitly described in the sixth chapter of Revelation.

The Sixth Seal earthquake may be an unusually large swarm of earthquakes. Japan had the largest recorded swarm of earthquakes.[6] It triggered a series of quakes to propagate through a group of volcanic islands from June 26, through August 26, 2000. Over 500 earthquakes with magnitudes greater than 4.0 were recorded. Five of them had magnitudes greater than 6.0 and the largest volcanic eruption occurred on August eighteenth. The great feature of the Sixth Seal earthquake may be a global swarm of earthquakes that relocate mountains and islands.

Sediments have been temporarily liquefied by earthquakes. Liquefaction has caused several architectural structures to briefly float, rotate, and precariously settle. Large seismic waves, immense geographical displacements, awesome atmospheric luminescence,[7] and expansive liquefaction have been associated with great earthquakes.[8] But, none of the historical ones satisfy the global characteristic of the Sixth Seal earthquake.

Section 1-5. Implications

The prophecy in Matthew 24:4-8 predicts a period that is called the beginning of sorrows. The Old English word sorrows was translated from the Greek word "odin" for birthpangs (the pattern of pains associated with childbirth). We now use the term contractions for birthpangs. According to Matthew, earthquakes were one of the forms of disasters that are to occur with increasing frequency and intensity – like contractions. Disaster tables in earlier almanacs seem to evince that prophesied trend. On the other hand, analyses performed by the United States Geological Survey (USGS) did not yield that trend. USGS researchers attribute the apparent increase in frequency and severity to the increased numbers of seismic research facilities. None of the earthquakes have caused the specific sequence of events described by Revelation 6:12.

The Sixth Seal earthquake is not the greatest earthquake. Revelation 16:18-20, a description of the outpouring of the seventh vial of wrath, will trigger the worst earthquake. It causes many cities to fall, all islands to flee away, and all mountains to disappear. Subsequently, Greenland, the world's largest island (840,000 square miles), would be completely removed by that disruption. No other earthquake will match its severity. The subsequent storms of hundred pound hailstones will reduce the technologies of mankind to 12th century, pre-gunpowder capabilities.

The energy of the Sixth Seal earthquake shall be many times larger than other strong earthquakes – every mountain and island are moved from their places! What kind of mechanism would drive the Sixth Seal earthquake to move mountains without the destruction of cities? According to the United States Geological Survey, magnitude 9.2 earthquakes are the upper limit of what rocks can generate. The mechanism that dislodges every mountain and island involves a different distribution of energy. The Chile earthquake was 9.6 and did not dislodge mountains.

Another ground shaking mechanism is volcanic explosion. The Tambora and Krakatau explosive eruptions shook a couple of mountains and darkened the images of both the Sun and Moon. If the Sixth Seal earthquake was generated by a volcanic eruption, a small ice age would be caused by the debris of fine particles suspended in the air. The sixth chapter does not describe wintry conditions.

Another mechanism that could cause earthquakes throughout the planet and dislodge mountains will be discussed in chapter six of this book. Chapter two will provide a couple examples from historical documents of earthquakes linked to sudden losses of sunlight. The goal of this chapter was to illustrate how normal seismic events, though very powerful, are limited to regional disturbances. They do not have the energy to cause global shifts in mountain positions.

Chapter 2
The Sun Becomes Black as Sackcloth of Hair

> And I beheld when he had opened the sixth seal, and, lo, there was a great earthquake; **and the sun became black as sackcloth of hair**, and the moon became as blood; (Revelation 6:12c)

Is it preposterous to believe the Sun will become black? Some authors have interpreted this passage as a description of clouds blocking sunlight. Ashes from volcanic eruptions, debris from meteor impacts, gamma ray induced smog from a supernova, and smoke from a nuclear war have been used to explain this verse. A few others have tried to describe it in terms of a rogue star colliding with the Sun. A shortcoming shared by all of those mechanisms is the absence of the element of surprise. Rogue stars, meteors, and missile warheads can be tracked before impact. And, clouds take time to form and cover vast areas.

Taylor had difficulty with the literal interpretation of this passage. He thought the word black, in the above passage, indicated the cessation of all light from the Sun. He assumed a literal solar blackout would cause all life to cease because photosynthesis would stop producing oxygen.[9] Taylor's deduction overlooked the possible cessation of only visible light. Revelation 7:1,3; 8:7; 9:4 contain descriptions of grass and trees beyond the sixth chapter.

Diminished production of visible light would not pose a grave threat to the plant kingdom. The absorption spectrum of chlorophyll and the action spectrum of photosynthesis have blue and red light as the most important sources of energy.[10] Chlorophyll is the principle component of photosynthesis. Cessation of visible light by the Sun would not terminate the plants' production of oxygen. Chlorophyll could get its energy for photosynthesis from the ultraviolet segment of the spectrum. Pollination would not be hindered by the lack of visible light because bees and birds can see the ultraviolet light markings on plants.

Darkening only one of the segments of the Sun's emission spectrum is not as preposterous as it sounds. Other segments of the Sun's spectrum darken periodically. The intensity of light from the soft X-ray region of the Sun's spectrum varies drastically with the sunspot cycle. Books and magazines have carried the series of photographs taken by the Yohkoh satellite of the Sun's corona in the X-ray region of the spectrum. They show the corona gradually darkening over several years in the soft X-ray band of the spectrum. The variations in the soft X-ray segment of the spectrum follow the eleven-year sunspot cycle.

It is not preposterous to believe the Sun will become black in the near future. Reports from the *Holy Bible* and other historical records have described incidents of solar blackouts that could not be attributed sunlight blocking phenomena such as eclipses and clouds. Those documents stated it has happened before and biblical prophecy has stated it will happen again.

One of the arguments for physically interpreting Revelation 6:12 stems from historical accounts of the Sun darkening. This chapter presents those records as evidence for rare occurrences of acute solar darkening events (ASDEs). Neither eclipses nor clouds have been effectively used to satisfactorily explain all of the details about an ASDE. Section 2-1 reviews the historical trustworthiness of the New Testament to justify treating all of the details of the crucifixion of Jesus Christ as physical realities. Section 2-2 presents the biblical and non-biblical accounts of the darkness and earthquake at the crucifixion of Jesus Christ. Section 2-3 surveys the variety of attempts to explain the crucifixion darkness. Section 2-4 reviews the rationale and hypothesis that can be used to identify records of ASDEs. Criteria for classifying abrupt losses of sunlight as an ASDE are given in Section 2-5. Section 2-6 summarizes the conditions that may have to occur to gain the recognition of ASDEs as a reality prior to the physical fulfillment of Revelation 6:12.

Section 2-1. Historical Trustworthiness of the New Testament

The *Holy Bible* is like a library.[11] Many of its volumes can be used to promote mankind's understanding of physical phenomena. Earlier works about the darkness at the crucifixion did not treat all of the biblical details as factual components of historical reality. Such actions were tantamount to disregarding one of the major cannons of historiography: assume the works of a historian are true.[12] Their arguments (1.) have stated the descriptions of totality were exaggerations and/or metaphors; (2.) have claimed it had not occurred because of the absence of accounts by prominent, non-Christian historians; and/or (3.) have tried to attribute the darkness to either an eclipse or clouds.

The historical reliability of the Gospels received additional confirmations from many quality works that were published during the last two decades of the twentieth century.[13,14] Those studies have weakened the assumptions the crucifixion darkness accounts had been the products of exaggerations and/or metaphorical expressions. Luke's works have proven to be one of the most reliable sources of historical facts about the second half of the first century.[15] Several historians and biblical critics have acknowledged the historical trustworthiness of Luke.[16] Subsequently, Luke's writings can serve as a source of detailed facts about the darkness at the crucifixion. Anderson's treatment of that phenomenon as a historical reality did not settle the question of its extent.[17] This chapter provides additional arguments for the global interpretation.

With respect to the second argument against the historical reality of the crucifixion darkness, the silence of non-Christian historians was actually self imposed. Upon considering the lack of attention paid by ancient Graeco-Roman historians to Palestinian figures, the self imposed censorship of non-Christian historians should be expected.[18] Weston's examination of the practices of Ælian, Aurelius Victor, Dio, Eutropius, Ovid, Pliny the Elder, Plutarch, Servius, Sextus Ruffus, Solinus, Suetonius, Tacitus, Tibullus, Valerius, Virgil, and Zosimus evinced their disdain of Christian miracles.[19] Roman leaders, the intended readership of early histories, were not interested in laudatory phenomena associated with a Jewish prophet.[20] In spite of the deteriorated relations between

Christians and Jews, the latter produced more writings about the life of Jesus than the Graeco-Roman historians.[21] They did not comment on the crucifixion darkness. The Egyptians were not among the set of crucifixion darkness writers. That was consistent with their practice of not recording solar eclipses.[22]

With respect to the third argument, Section 2-3, Explanatory Attempts, provided the salient astrophysics principles that invalidated ecliptical explanations of the crucifixion darkness. Non-biblical documents attesting to the cloudless and clear sky conditions were presented in Section 2-2 and 2-5.

Section 2-2. Crucifixion Darkness Accounts

One of the most firmly established details about the darkness at the crucifixion is the time of day it had started. With respect to historical authenticity, the absence of discrepancies between sources about a detail allows it be treated as an unimpeachable fact.[23] The Synoptic Gospels (Matthew 27:45; Mark 15: 33; Luke 23: 44-45), Christian apocrypha (*Gospel of Peter*,[24] *Gospel of Nicodemus*,[25] and *The Acts of John*[26]), Phlegon's *Olympiades*,[27] and Orosius' *Seven Books of History Against the Pagans*,[28] stated the darkness had commenced at the sixth hour.

The primary cause of excellent agreement between the diverse sources may have been the widespread disruption of customary activities that had been caused by the darkness. Judaism had required prayers three times daily and Islam, five.[29] The midday prayers would have transpired close to the sixth hour. For example, Peter had followed the noonday custom (Acts 10:9). In accordance with Job's prophecy (Job 5:13-14), the unexpected onslaught of darkness at noon stopped the mockery by the chief priests, elders, scribes, and others that had been directed against Jesus as He hung on the cross.

Phlegon provided a description of a darkening that had been accompanied with an earthquake. He attributed the darkness to a solar eclipse. Various writers have claimed Phlegon had dated it to the 203rd Olympiad; the 2nd year of the 102nd Olympiad; the 2nd year of the 202nd Olympiad; or the 4th year of the 202nd Olympiad.[30,31] The discrepancies were resolved by Eusebius' verbatim quotation of Phlegon's *Olympiades*: "In the 4th year of the 202nd Olympiad there happened an eclipse of the

Sun, greater than any that ever had been known before: and there was such darkness at the sixth hour of the day that the stars appeared in the heavens. It was accompanied likewise with a great earthquake in Bithynia which overthrew a considerable part of the city of Nice."[32] Later English translations spelled the city's name as Nicæa and had pluralized the word 'earthquake.' Attempts to associate Phlegon's darkening with the crucifixion darkness had been justified because the 4th year of the 202nd Olympiad had transpired from A.D. 32 July 1 through A.D. 33 June 30.[33] That period accommodated A.D. 33 as one of the many dates calculated for the crucifixion. Fotheringham had assumed Nicæa, Bythinia, to have been the site of Phlegon's report.[34]

Several methods had been used to determine the dates of the crucifixion. Finegan summarized the variety of historical chronologies that have been used to determine the dates of the Passion.[35] Applications of the seventy weeks prophecy of Daniel to the date of Artaxerxes' order to rebuild Jerusalem have yielded A.D. 32 April 6[36] and A.D. 33 March 30[37] for Palm Sunday. Stockwell, in 1892, obtained Friday, A.D. 33 April 3 based on the B.C. 6 May 8 conjunction of Jupiter and Venus, as the star of Bethlehem, and the paschal full moon falling on a Friday.[38] Other astronomical methods have been based on the date of the new moon – the first day of Nisan. Passover occurs in the middle of the month during a full moon.[39] Sir Isaac Newton, in 1733, was the first to derive a date (Friday, A.D. 34 April 23) by determining when the crescent of the new moon became visible.[40] The two accepted dates by the scientific community, based on such calendrical determinations, have been Friday, A.D. 30 April 7, and Friday, A.D. 33 April 3 – with higher recognition going to the latter.[41]

The global extent and non-ecliptical characteristic of the darkness was stated in A.D. 198 by Tertullian, in his greatest work, *The Book of Apology Against the Heathen*: "At the same moment the light of midday was withdrawn, the sun veiling his orb. They thought it forsooth an eclipse, who knew not that this also had been foretold concerning Christ: when they discovered not its cause, they denied it; and yet ye have this event, that befel (sic) the world, related in your records."[42] Note the phrase "when they discovered not its cause" means they realized it could not be attributed to an eclipse and/or clouds. A lengthy footnote for the word 'records' referred to an account probably sent by Pilate to the archives and the existence of additional Greek memoirs mentioned by Eusebius that were independent of Phlegon and Thallus. Tertullian and Lucian had declared the secret archives of the empire, then in existence, possessed reports that evinced the supernatural darkness that had prevailed during the death of Christ.[43,44] Ussher recorded the following statement by Lucian: "Search your writings and you shall find that, in Pilate's time, when Christ suffered, the sun was suddenly withdrawn and a darkness followed."[45] According to Weston, proconsuls, prefects, and other authorities were required to notify the Roman senate of any abnormal events that had occurred within their provinces.[46]

The footnote for the word 'foretold,' in Dodgson's translation of Tertullian's work, referred to Amos' prophecy: "And it shall come to pass in that day, saith the Lord God, that I will cause the sun to go down at noon, and I will darken the earth in the clear day" (Amos 8:9). Phlegon's note about the stars becoming visible attested to the clear sky conditions.

Matthew's writings stated it had covered all of the land (Matthew 27:45). Mark said it had covered the whole land (Mark 15:33). Luke stated it had covered all of the earth (Luke 23:44). Even though Luke's trustworthiness as a historian had been acknowledged by historians and biblical critics,[47] his statement about the globally darkened Earth probably stemmed from a divine revelation (2 Timothy 3:16).

The Synoptic Gospels stated the darkness stopped at the ninth hour. Witnesses did not need a sundial or hourglass to know when the sixth and ninth hours had occurred. Ancient cultures kept time by pointing to specific positions of the Sun.[48] Palestine, like many Mediterranean nations, had to measure time in terms of the Roman twelve divisions of daylight: hours. According to Duncan,[49] the Roman soldiers announced the third hour of the morning (*tertia hora*), the sixth of midday (*sexta hora*), and the ninth of the afternoon (*nona hora*). During the month of the Passion this would be equivalent to 9:00 AM, 12:00 Noon, and 3:00 PM. General references to time were expressed in terms of quarter-days.[50,51]

According to the oldest, extant un-canonical description of the Passion, the apocryphal *Gospel of Peter*, people thought night had returned, had obtained lamps, and had went about, stumbling.[52]

All Shall Hide

According to Kidger, there would not have been enough time to light lamps if the darkness had been caused by a solar eclipse.[53]

Luke 23:44-45 (KJV) stated the cause of the darkness: "And it was about the sixth hour, and there was a darkness over all the earth until the ninth hour. And the sun was darkened, and the veil of the temple was rent in the midst." The *Codex Vaticanus*, *Codex Sinaiticus*, and other translations used the phrase "the sun's light failed" while the *Codex Alexandrinus* and other documents have "the sun darkened."[54] The phrase "the sun's light failed" was equivalent to "the sun was darkened." The Greek word for failed in the above appeared in an earlier passage by Luke 22:32 about weakening faith: "But I have prayed for thee, that thy faith fail not: and when thou art converted, strengthen thy brethren."

Orosius, in A.D. 418, provided an explicit account of the crucifixion phenomena in his tremendously successful *History Against the Pagans*.[55,56] It was consistent with Matthew's, Luke's, and Phlegon's writings. Book VII, chapter 4 of Orosius' indicated the darkness of the crucifixion had suddenly started at the sixth hour; could not be attributed to the clouds; indicated the Moon had not been in position to block the sunlight; the stars were visible; it had been accompanied with a great earthquake that shook the world, had fractured rocks atop of mountains, had severely damaged many cities in Asia; and both the Holy Gospels and a few books by the Greeks had recorded the phenomena.[57]

The *Codex Vaticanus*, *Codex Sinaiticus*, and *Codex Alexandrinus* used the Greek phrase ΚΟΤΟCΕΓΕΝΕΤΟ (darkness it became) to describe the darkness in the Gospels.[58] Pollata argued it implied the onslaught of darkness had been too rapid for a solar eclipse.[59] Although most English versions of the *Holy Bible* have not implied a rate of darkening, Pollata's observation is consistent with Orosius' report.

After the sunlight had returned at the ninth hour Jesus gave up the ghost. A moment later, the ground shook. Matthew, Mark, and Luke indicated the veil of the Temple was rent, but Matthew associated it with an earthquake, rending of rocks, and the opening of the graves of saints. Matthew's skills in tachygraphy (a form of shorthand) had enabled him to provide comprehensive descriptions and to record lengthy sermons verbatim.[60] Phlegon said the earthquake had rocked Nicæe. *The History Against the Pagans*,[61] and the apocryphal works *Gospel of Peter*[62] and *Gospel of Nicodemus*,[63] described it in global terms.

Matthew 28:2 stated another earthquake accompanied the resurrection a couple of days later.[64]

The accounts of seismic activity were historical facts. Phlegon's report was corroborated by Ambraseys.[65] He cited documents about the destruction suffered by cities throughout Asia (Turkey). Ambraseys stated the seismic phenomena at the crucifixion and resurrection had not been fault driven.[66] He suggested the Jerusalem reports of the crucifixion and resurrection quakes were exaggerations because of the lack of geological evidence and absence of civil incidence reports. His report did not explicitly address the rock of Golgotha. The Stone of Calvary (Golgotha), located in the Chapel of Adam at the Church of the Holy Sepulchre,[67] contains an extensive crack that has been attributed to the earthquake at the crucifixion.[68] According to Trench, Cyril of Jerusalem, the revered theologian of the Roman Catholic Church, the Eastern Orthodox Church, and the Anglican Communion, had declared (*c* 380 AD): "To this day Golgotha shows the spot where the rocks were rent because of Christ."[69] The Stone of Calvary evinced Matthew's and Orosius' accounts of stones fracturing atop mountains. The seismic shocks at Jerusalem that accompanied the crucifixion and resurrection were real.

Each critical stage of the Passion had transpired on extremely significant anniversaries and hours of the day. The *Holy Bible*, *Talmud*, and other sources indicated the crucifixion of Jesus had occurred during the eve of or during the early segment of Passover.[70] The Passover had been created during the preparations for the exodus from Egypt. God had commanded Moses and Aaron to have the congregation of Israel to get from the flocks on the tenth day of Nisan (March/April) one year-old male lambkins. They were to sacrifice them on the fourteenth day of Nisan. The lambkins' blood was to be spread on the two sides and upper door posts of homes. This would cause the plague of the death of the first-born to pass over those households. God commanded them to keep the Passover forever (Exodus 12:1-14). Jesus Christ, also referred to as the Lamb of God, was sacrificed for the sins of the world on the anniversary of Israel's deliverance from bondage. Jesus had entered Jerusalem a few days before Passover on Palm Sunday, the tenth day of Nisan. The following Sunday morning, an earthquake accompanied the resurrection of Jesus Christ. It fell on the seventeenth of Nisan -

the anniversary of Noah's ark coming to rest within the mountains of Ararat.[71]

The state of the Sun during the crucifixion was not tranquil. Calculated dates for the crucifixion fell within the solar activity period known as the Roman Maximum (BC 20 to A.D. 80).[72] This suggests solar magnetic storms may be conducive to the sudden solar blackout. We shall explore this thought further in Sections 2-4, 2-5, and 2-6.

Section 2-3. Explanatory Attempts

Miraculous Eclipse. The physical causes of a solar eclipse were understood in ancient times. Bartlett cited several references and diagrams to prove officials during ancient times and the Middle Ages knew the Moon's shadow caused solar eclipses and the Earth's shadow caused lunar eclipses.[73] *Tractatus de spera*, the Middle Age astronomy text used by universities for hundreds of years contained the following assessment of the crucifixion darkness: "… that eclipse was not natural but, rather, miraculous and contrary to nature."[74] That statement was equivalent to saying God had miraculously relocated the relative positions of the Sun and Moon to generate the crucifixion darkness. Descriptions of the solar darkening, attributed to Dionysius the Areopagite, Paul's convert from Athens (Acts 17:34), were in accord with the miraculous eclipse explanation. But, his alleged exclamations were not in the original Dionysian writings[75] and have been treated as the writings of Pseudo-Dionysius. Accounts of the crucifixion darkness by Luke, Tertullian, and Orosius did not include descriptions of drastic changes in lunar motions.

Solar Eclipse. The crucifixion darkness could not be attributed to an eclipse because the reported length of totality exceeded the maximum duration for total eclipses: 7 minutes 31.1 seconds.[76] The following are two well known attempts that had been made to attribute the crucifixion darkness to an eclipse:

Thallus had written a history of Greece's relationships with Asia in A.D. 52. Although none of those writings had survived, Julius Africanus, in A.D. 221, made the following statement regarding his explanation for the crucifixion darkness: "Thallus, in the third book of his histories, explains away this darkness as an eclipse of the sun unreasonably, as it seems to me."[77] Africanus' subsequent texts explained the impossibility of a Passover solar eclipse.[78,79]

The solar eclipse of A.D. 29 November 24 had been used by several papers to explain the crucifixion darkness. According to Chambers: "This is the only solar eclipse which was visible at Jerusalem during the period usually fixed for Christ's public ministry. This eclipse was for a long time, and by various writers, associated with the darkness which prevailed at Jerusalem on the day of our Lord's Crucifixion, but there seems no warrant whatever for associating the two events."[80] Fotheringham stated: "In spite of the date given by Phlegon, it is impossible to identify this eclipse with any except that of +29 November 24, which fell in the first, not in the fourth, year of the 202nd Olympiad."[81] That eclipse had not been dark enough to be noticeable in Jerusalem by those outdoors.[82,83]

Sawyer assumed every verse in the Gospels had not been a source of factual information to argue Luke's account was a reminiscence of the 29 November 24 eclipse.[84] He argued the Greek root for the darkening should have been translated as eclipse. Most English translations of Luke's statement have not attributed the darkness to a solar eclipse. The NKJV, YNG, DBY, WEB, and the Hebrew Names Version use very similar wording to indicate the Sun had been darkened. The ASV, ESV, and RSV implied the ability of the sunlight to shine had failed. The NASB stated the sunlight was blocked. The NIV implied the Sun abruptly failed to emit light. And, the NLT indicated the Sun disappeared.

Some have erroneously claimed Amos' prophecy for the darkness at the crucifixion had been fulfilled by a total solar eclipse. According to the *Holy Bible*:

> And it shall come to pass in that day, saith the Lord God, that I will cause the sun to go down at noon, and I will darken the earth in the clear day: (Amos 8:9)

For example, Brewer had stated: "'That day' was June 15, 763 B.C." and it had been recorded by a scribe in Nineveh.[85] Translations of that record did not identify the time of day of that famous eclipse.[86] It is known as the Eponym Canon Eclipse; it had been recorded in the *Assyrian Chronicle*; and has been recognized as the earliest record

All Shall Hide

of a solar eclipse. Fotheringham indicated it may have been total over Nineveh, where the extant copies of the *Assyrian Chronicle* had been discovered or at Calah, the capital of Assyria during the time of the eclipse.[87] Fotheringham was satisfied with locating it somewhere in Assyria. Stephenson asserted the lack of details from the *Assyrian Chronicle*, could have placed it at the capitol or at one of its provinces.[88] Hind calculated the eclipse would have passed south of Nineveh; 4 minutes 20 seconds would have been its maximum length of totality; and "half past 9 local time" as the middle of the eclipse.[89] According to Hind, it would have appeared as a large partial solar eclipse at Nineveh. Johnson calculation of 9:47 AM for its totality near Nineveh was in agreement with Hind's calculation.[90] Neither Hind's nor Johnston's determination of the time of the eclipse, mid-morning, corresponded to the "noon" specified by Amos. The physical details of the Eponym Canon Eclipse and conventional obscuration mechanism have not been able to satisfy the specific criteria of Amos 8:9. Note, that scripture indicates the Sun will be turned down – like the volume of a radio – during a cloudless day.

Lunar Eclipse. Humphreys and Waddington had argued the reddened Moon from a lunar eclipse had inspired the apostles to write of the crucifixion in terms of Joel's prophecy.[91] Schaefer proved the eclipsed Moon would not have been visible at moonrise,[92,93] and could not have a "blood color" after moonrise due to optically masking effects of the penumbra..[94]

The crucifixion darkness was not a fulfillment of Joel's prophecy. Luke provided a further account, in Acts, that placed the crucifixion darkness beyond Joel's prophesized sequence of events that culminated with the combination of a blackened sun and redden moon. Seven weeks after the resurrection, during the Day of Pentecost, the hundred and twenty were filled with the Holy Ghost (Acts 2:4). People in the streets began to attribute their peculiar behavior to drunkenness. Luke recorded Peter's explanation in Acts 2:12-20:

> And they were all amazed, and were in doubt, saying one to another, What meaneth this? Others mocking said, These men are full of new wine. But Peter, standing up with the eleven, lifted up his voice, and said unto them,

Ye men of Judaea, and all ye that dwell at Jerusalem, be this known unto you, and hearken to my words: For these are not drunken, as ye suppose, seeing it is but the third hour of the day. But this is that which was spoken by the prophet Joel; And it shall come to pass in the last days, saith God, I will pour out of my Spirit upon all flesh: and your sons and your daughters shall prophesy, and your young men shall see visions, and your old men shall dream dreams: And on my servants and on my handmaidens I will pour out in those days of my Spirit; and they shall prophesy: And I will shew wonders in heaven above, and signs in the earth beneath; blood, and fire, and vapour of smoke: The sun shall be turned into darkness, and the moon into blood, before that great and notable day of the Lord come.

Darkness at the crucifixion could not apply to Joel's prophecy because its predicted intervening events (e.g., wonders in heavens and signs in the earth beneath and vapour of smoke) had not transpired. Walvoord,[95] Lockyer,[96] Hoyt,[97] and Bloomfield[98] had argued the crucifixion darkness had not been a fulfillment of the scripture in Joel.

Physical Eschatology. An acute solar dimming explanation for the crucifixion darkness has not been recognized by the solar evolution theories of the scientific community. They have predicted several billion years will have to elapse before the Sun disappears from view. The Sun is expected to expand to become a red giant; contract to a helium core-burning star; expand to a double shell-burning red giant; expel its outer layers as planetary nebula to leave a white dwarf; and then, after an indefinite period, cool to become a black dwarf.[99]

Section 2-4. Acute Solar Darkening Events Hypothesis

What can be learned from the accounts of the darkness at the crucifixion? Neither obscurations (clouds and/or eclipses) nor perturbations (e.g., collision with a dense, rogue mass) had caused it. The historical records suggest it had been generated by an internal process. It shall be referred to in this book as an acute solar darkening event (ASDE). A hypothesis was formed from the crucifixion darkness records that can improve our understanding of the relationships between the varieties of phenomena described by the pre-wrath scriptures. The four hypothesized attributes of an ASDE are night-like darkness, reoccurrences, accompanying intense solar activity, and ASDEs in sun-like stars. Evidence for ASDEs was acquired by applying the selection criteria given in the hypotheses to historical documents and eclipse reports. The subsequent findings were used as additional arguments for physically interpreting Revelation 6:12-17.

The first criterion of the ASDE hypothesis involved the anomalously large reductions in brightness. An ASDE is much darker than a sunspot. Even though sunspots appear dark relative to the rest of the Sun's surface, they are as bright as a welder's arc.[100] This can be numerically expressed in terms of apparent magnitude. Apparent magnitude is a form of measurement of brightness. The more negative the apparent magnitude, the brighter the object. The larger the positive value of an apparent magnitude, the darker the object. The apparent magnitude of the Sun is -26.7. The apparent magnitude of the faintest star visible to the naked eye, under perfect, night viewing conditions, is +6.5. The Sun would have to change its apparent magnitude by 33.2 to attain the darkness at the crucifixion. That change is equivalent to reducing its brightness down to 5.2×10^{-14} of its normal value. Sunspots have apparent magnitudes between -15 and -14.5. If the entire Sun was covered with sunspots, it would be brighter than the Moon. The full moon has a -12.5 apparent magnitude.[101] The R Coronae Borealis (R CrB) type star is the only class of irregular variable stars that could approach ASDE behavior.[102] Dimming light curves of R CrB stars yield changes in apparent magnitude between 6 and 7 – values that are much smaller than the 33.2 change in magnitude for an ASDE.

Darkness, the first attribute of the ASDE hypothesis applies only to the visible segment of the solar spectrum. All frequencies of light are blocked by the Moon during total solar eclipses. This attribute of the hypothesis allows emission levels of the invisible portions of the Sun's spectrum to vary in intensity during an ASDE.

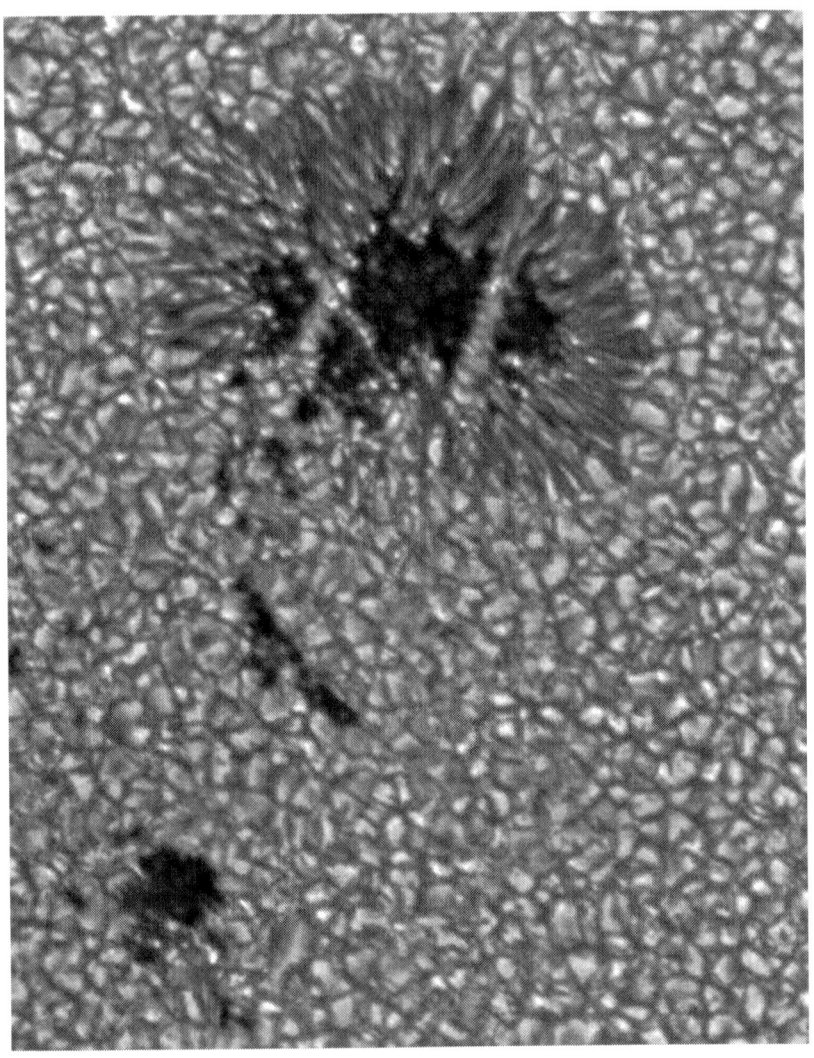

Figure 1. The Sun will be much darker than sunspots during the opening of the Sixth Seal.

All Shall Hide

The second criterion of the ASDE hypothesis states they reoccur. Solomon summarized that property with the following aphorism: "The thing that hath been, it is that which shall be; and that which is done is that which shall be done: and there is no new thing under the sun. Is there any thing whereof it may be said, See, this is new? it hath been already of old time, which was before us." (Ecclesiastes 1:9). This scripture was used as a justification for searching for historical accounts and predictions for future ASDEs.

Non-biblical and biblical sources have been available for testing the second attribute of the ASDE hypothesis. Chronicles and their analyses were used as the non-biblical sources that were examined for accounts of past ASDEs supplementing the crucifixion darkness. Joel 2:31; Acts 2:20; Revelation 6:12, 8:12; Matthew 24:29; and Mark 13:24 were designated as biblical sources to be examined for predictions of future ASDEs.

The ASDE reoccurrence tests were literature searches for multiple, independent, reputable accounts of total solar eclipses with (1.) dates and locations impossible for a total solar eclipse; and/or (2.) lengths of totality exceeding a half hour. The latter criterion stemmed from the 7 minutes 31.1 seconds maximum length of totality for total solar eclipses.[103]

Works by F. Richard Stephenson and a variety of major chronicles were the historical sources examined for evidence of ASDEs. Stephenson's *Historical Eclipses and Earth's Rotation* has been an excellent compilation of analyses of historical observations of lunar and solar eclipses prior to A.D. 1600. It is a culmination of thirty years of research into eclipses pertinent to determining changes in the Earth's rotation.[104] Analyses by Stephenson and Yau of thirty-seven solar eclipses recorded in the *Ch'un-ch'iu Chronicle*, from B.C. 720 to B.C. 481, was another source.[105] A third source was the collection of annals known as the *Anglo-Saxon Chronicle*. It was composed of material dated from B.C. 60 to A.D. 1154.[106] The fourth category of documents were the catalogues of pre-telescopic sunspot observations compiled by Clark and Stephenson;[107] Willis, Armstrong, Ault, and Stephenson;[108] Willis and Stephenson;[109] and Yau, and Stephenson.[110]

The third criterion of the ASDE hypothesis states they transpire during periods of intense solar activity. This attribute was based on correlations between earthquakes and sunspot activity that have been manifested. Mazzarella and Palumbo established a correlation between the eleven-year sunspot cycle and large earthquakes.[111] Their analyses examined the solar and geomagnetic phenomena associated with Italian earthquakes that occurred from 1833 through 1930. The total absence of large earthquakes during the seventy-year period of relative solar calm, known as the Maunder Minimum (1645 – 1715), was one of the interesting observations presented in their study. Palumbo found similar correlations between solar and seismic activity for world-wide, West Pacific, and Italian earthquakes.[112] Shaltout, Tadros, and Mesiha examined data on worldwide earthquakes of $M \geq 5$ and sunspot numbers for the period of 1903-1985.[113] Their analyses manifested a significant, but not exclusive, role of the Sun as a trigger of earthquakes. The top two from their list of seven physical catalysts were "(1) Abrupt accelerations in the Earth's angular velocity (the daily speed of rotation of the earth)" and "(2) Surges of telluric currents in the Earth's crust." Analyses of worldwide, Richter scale 6 earthquakes from 1958 through 1988 by Shatashvili, Sikharulidze, and Khazaradze yielded a correlation with the eleven-year sunspot cycle.[114] They distilled a 75% occurrence rate for magnitude 6 quakes within a day whenever the Earth crosses the neutral sheets of the interplanetary magnetic field. Further analyses yielded correlations between sunspot activity and earthquakes. An examination by Han, Guo, Wu, and Ma of twenty-two earthquakes in China, from 1556 to 2001, yielded a correlation between solar activity and earthquakes.[115] They found the dates of large earthquakes near faults with west-east strikes occur close to dates the sunspot numbers had reached their maximum.

The growing evidence for complex connections between solar activities and earthquakes justified forming a hypothesis that stated the existence of a causal relationship between the darkening of the Sun and subsequent seismic disturbances that accompanied the crucifixion. This attribute of the ASDE hypothesis does not claim earthquakes accompany every solar darkening. Comparisons with the dates of the crucifixion darkness and ASDEs and the (1.) periods of solar activity maximums and (2.) irradiance levels of the Perry and Hsu solar output model were used as the tests.

The irradiance levels of the Perry and Hsu solar output model[116] were used to facilitate comparisons between various ASDEs and levels of solar activity. For example, the Perry-Hsu solar output model[117] yielded 8.02% for the irradiance level of the Sun during A.D. 33, the year of the crucifixion of Jesus Christ. The model gives 0.45% for the irradiance level during A.D. 1648 of the Maunder Minimum.

The fourth criterion of the ASDE hypothesis indicates sun-like stars should have similar phenomena: acute celestial darkening events (ACDEs). This condition stems from applying Ecclesiastes 1:9 to Revelation 8:12: "And the fourth angel sounded, and the third part of the sun was smitten, and the third part of the moon, and the third part of the stars; so as the third part of them was darkened, and the day shone not for a third part of it, and the night likewise." Such expectations are consistent with the Mediocrity Principle. Earth-based observations of ACDEs are likely to be attributed to either irregular variable stars or anomalous meteorological phenomena.

The ability to detect and reliably document contemporary ACDEs has been enhanced by the launch of NASA's Kepler Mission.[118] It will search three years for terrestrial planets by continuously monitoring over 100,000 stars. The Kepler Spacecraft shall look for faint drops in starlight caused by the orbits of planets crossing a star as seen from Earth.[119] Expected transit durations range from 30 minutes to half a day and ACDE decreases in starlight (99.99%), would exceed the expected faint dips of the Mission (<1%).[120]

The remainder of this chapter will focus on the historical evidence for the first three attributes of the ASDE hypothesis.

Section 2-5. Test of Hypothesis

Solar eclipse records contained more evidence for ASDEs than the sunspot compilations. Only one potential candidate was identified among the sunspot catalogues. Clark and Stephenson[121] cited an AD 1597 June 15 report of the blackening of the solar disk that had been regarded by the Yunnan Observatory as 'dubious.' Scuderi indicated that phenomenon may be attributed to atmospheric obscuration of the Sun by the 1597 eruption of the Icelandic volcano Hekla.[122] The Global Volcanism Program web site indicated the Hekla eruption had started January 3rd and ended by or in June of that year.[123] Iceland is located in the North Atlantic Ocean, several thousand miles away from China. A significant solar darkening caused by Hekla would have been an atmospheric anomaly! The June 15, 1597 report of the blackened Sun could not be attributed to an eclipse. According to the NASA Eclipse Web Site, only two annular solar eclipses occurred in 1597 (March 17th and September 11th). The AD 1597 June 15 report was excluded from the evidence for ASDEs because of the absence of multiple independent sightings – not because of the Hekla eruption.

Twelve findings were generated from the tests for ASDE reoccurrences to supplement the crucifixion ASDE. The following accounts satisfy the first two criteria of the ASDE hypothesis:

(1.) B.C. 645 April 03. Stephenson's examination of five total solar eclipses (B.C. 691 to 637) and eight annular solar eclipses (B.C. 689 to 633) proved they were not dark enough in either the Aegean Sea islands of Thasos or Paros nor at Sparta to account for the sudden and intense darkening described by the poet Archilochus.[124] A writer in Ch'ü-fu, China, had recorded in the *Ch'un-ch'iu Chronicle* a solar eclipse for the date of B.C. 645 April 03. It was attributed to meteorological phenomena because an eclipse was impossible for that region for more than a year before and after that date.[125] Since the period of Archilochus' poem and the date of the Chinese observation could not be attributed to a total solar eclipse, they can be treated as evidence for an ASDE. With respect to solar activity periods, B.C. 645 April 03 transpired five years before the 4th grand maxima of the SN-S series (B.C. 640 - 550).[126] The corresponding irradiance level for B.C. 645, according to the modified Perry-Hsu solar activity model, is 24.43%.

(2.) B.C. 592 May 15. The abrupt onslaught of darkness had stopped a war between the Lydians and the Medes. Chambers provided the following description by Herodotus:

> As the balance had not inclined in favour (sic) of either nation, another engagement took place in the sixth year of the war, in the course of which, just as the battle was growing warm, day was suddenly turned into night. This event had been foretold to the Ionians by Thales of Miletus, who predicted for it the very year in which it actually took place. When the Lydians and Medes observed the change they ceased fighting, and were alike anxious to conclude peace.[127]

Many had assumed that darkness event could be attributed to a fulfilled prediction by Thales that had been recorded by Pliny despite the fact he had not mentioned the battle.[128] Pliny stated Thales of Miletus, during the 4th year of the 48th Olympiad, had predicted a total solar eclipse that occurred later in that year - the 170th year after the foundation of Rome.[129] Eclipses that had occurred in B.C. 610, 593, and 585 had been considered, but Airy's exhaustive work resolved the discussions to B.C. 585.[130] Stephensen indicated that eclipse (B.C. 585 May 28) and those of B.C. 588 and 581 had not been dark enough to satisfy the description by Herodotus.[131]

According to Fotheringham, Eusebius had not linked Herodotus' description of the battle with Pliny's account of the eclipse.[132] If Herodotus and Pliny had described the results of two different predictions by Thales, Herodotus' account could be attributed to an ASDE. A chronicler in Ch'ü-fu, China, recorded in the *Ch'un-ch'iu Chronicle* the solar darkening of B.C. 592 May 15 as an eclipse.[133] Although that chronicle has proven to be reliable, that date, like the one for B.C. 645, could not be associated with a total solar eclipse and has been attributed to meteorological phenomena. Therefore, Herodotus' description of the terminated battle between the Lydians and Medes and the *Ch'un-ch'iu Chronicle* record could be used as evidence of another ASDE. With respect to periods of solar activity, B.C. 592 May 15 falls within the 4th grand maxima of the SN-S series (B.C. 640 - 550).[134] The corresponding irradiance level for B.C. 592, according to the modified Perry-Hsu solar activity model, is 18.13%.

(3.) *B.C. 480 Early Spring.* Another darkening of the Sun that could not be attributed to a solar eclipse had been recorded by the "Father of History," Herodotus, an Athenian statesman and general, Aristides (Aristeides son of Lysimachos), and the great Greek biographer, Plutarch. They stated a sudden darkness had occurred when the Persians started their third invasion of Greece. Chambers' review of explanatory attempts included the following three accounts under the assumption they had described the same event:[135]

> Herodotus (*The Histories, VII, 37*): At the first approach of Spring the army quitted Sardis and marched towards Abydos; at the moment of its departure the Sun suddenly

quitted its place in the heavens and disappeared, though there were no clouds in sight and the day was quite clear; day was thus turned into night.

Aristides (Scholia, in Aristidis Orationes): As the king was going against Greece, and had come into the region of the Hellespont, there happened an eclipse of the Sun in the East; this sign portended to him his defeat, for the Sun was eclipsed in the region of its rising, and Xerxes was also marching from that quarter.

Plutarch (Life of Pelopidas): An army was soon got ready, but as the general was on the point of marching, the Sun began to be eclipsed, and the city was covered with darkness in the daytime.

The year of the Persian disembarkation, B.C. 480, had been firmly fixed in history. From B.C. 484 through 481, Xerxes had amassed an army at Sardis and a screening naval force to historically large levels.[136] The Persian advance commenced during the early Spring of B.C. 480. A few months later that year was the martyrdom of Leonidas and his 300 Spartans in the delaying action at the Battle of Thermopylae. A month later, the Persian naval armada suffered a resounding defeat at the Battle of Salamis. The subsequent victory of the Greeks against the Persian army at the Battle of Plataia transpired the following year. Thus the course of western civilization had been changed by the amazing victories of the Spartans and Athenians over the immense forces of the Persians. The dates for those significant events had been firmly established and there were no valid reasons for changing them.

Over four centuries of astronomers had proven eclipses could not satisfy the conditions described by Herodotus. Subsequently, some astronomers had suggested he had erroneously recorded the phenomena. The eclipse of B.C. 480 April 19 was not suitable because the shadow of the Moon had swept across the southern regions of India and China. The B.C. 481 April 19 was visible in Persia, not in Greece and Asia Minor. It, like the one in B.C. 480, fell on April – a month that does not satisfy the "first approach of spring" condition.[137] Hind[138] had considered the one seen by Cleombrotus on B.C. 479 October 2 to be the eclipse recorded by Herodotus (IX, 10)

[139] that had occurred after the Battle of Salamis in B.C. 480 September 20. Hind[140] and Lynn[141] indicated the best candidate was the annular solar eclipse (94% coverage) of B.C. 478 February 17 because it had been visible in the morning over Sardis. Chambers' summary of the problem included Airy's assertion that Herodotus had confused the account of the solar darkening with the lunar eclipse of B.C. 476 March 13.[142] Correspondingly, Stephenson asserted none of the eclipses were a suitable explanation and suggested Herodotus' record was another error.[143]

The famous 16th century, Danish astronomer, Tycho Brahe, accepted the historical accounts as descriptions of the Sun darkening without the assistance of the Moon. Brahe's assessment was stated in the following manner in his *Historia Celestis*: "Xerxes crossed over into Greece this year, as spring drew on. At this time, Herodotus asserts the sun was darkened. But this must have happened without an eclipse, as there was none in the spring of this year, or the former."[144] Herodotus, Aristides, and Plutarch had described another ASDE. The Sun had been darkened during a calm and cloudless day in B.C. 480. With respect to solar activity periods, B.C. 480 falls within the 3rd grand maxima of the SN-S series (B.C. 490 - 420).[145] The corresponding irradiance level for B.C. 480, according to the modified Perry-Hsu solar activity model, was 22.54%.

(4.) *B.C. 34 August 23.* An eclipse was recorded in Ch'ang-an, China, by chroniclers for the *Han-shu* and the *Basic Annals* as transpiring on the non-eclipse date of B.C. 34 August 23.[146] Since Stephenson had categorized the reliability of the multiple reports as Class B, these could be considered as evidence for another ASDE. The class B category was assigned to brief accounts of eclipses that have additional details provided by other sources. This book will treat Class B reports as the products of independent accounts that could serve as evidence for an ASDE. With respect to solar activity periods, B.C. 34 August 23 had transpired fourteen years before the Roman Maximum (B.C. 20 – A.D. 80).[147] The corresponding irradiance level for B.C. 34, according to the modified Perry-Hsu solar activity model, is 9.80%.

(5.) *A.D. 33 April 03.* The ASDE of the crucifixion of Jesus Christ, as described earlier, transpired during the Roman Maximum (B.C. 20 – A.D. 80). The corresponding irradiance level for A.D. 33, according to the modified Perry-Hsu solar activity model, is 8.02%.

(6.) *A.D. 879 April 25.* A well known darkening record that has been classified as erroneous was the brief description of the one-hour eclipse recorded in the *Anglo-Saxon Chronicle*. The chronicler cited only the length of totality and the year, A.D. 879, not the date.[148] Chambers cited a source that indicated it had started in the afternoon at 3:00.[149] Also, he stated a solar eclipse would not pass through London between the years of A.D. 878 and 1715.[150] This record is normally treated as an erroneous dating for the eclipse of A.D. 878 October 29. A similar claim originated in Ch'ang-an, China, in the Astrological Treatise of the *Hsin-t'ang-shu*, of a solar eclipse dated A.D. 879 April 25.[151] Stephenson attributed it to a false sighting recorded as an observation.[152] These are either coincidental errors or evidence for an ASDE. With respect to solar activity periods, A.D. 879 occurred 79 years after the Dark Age Minimum (A.D. 600 – 800).[153] The corresponding irradiance level for A.D. 879, according to the modified Perry-Hsu solar activity model, is 16.61%.

(7.) *A.D. 977 December 13*. Stephenson indicated the Annals and Astrological Treatise of the *Sung-shih* had recorded a total solar eclipse for this date that could not have been seen in China.[154] This Class B report was attributed to a prediction that had been reported as an observation. It should be treated as another form of evidence for an ASDE. The year A.D. 977 occurred between maximum and minimum solar activity periods. The corresponding irradiance level for A.D. 977, according to the modified Perry-Hsu solar activity model, is 12.28%.

The next four dates for possible ASDEs fall within the solar activity period known as the Medieval Maximum (A.D. 1120 - 1280).[155] Their proximity to each other may be a product of interdependence.

(8.) *A.D. 1133 August 02*. According to Stephenson, an eclipse had been reported to have started at midday to last an hour in both Kerkrade, Netherlands,[156] and Augsburg, Germany;[157] had started at the 9th hour in Heilsbronn, Germany;[158] had suddenly darkened at midday to last half an hour in Zwiefalten, Germany;[159] had started between the 7th and 8th hour to last a half hour in Reichersberg, Austria;[160] and the Sun had suddenly disappeared at midday in Salzburg, Austria.[161] According to Chambers' research, it had started midday and was accompanied with an earthquake in England.[162] Although a total solar eclipse had transpired on this date, its totality was much smaller than the durations given in the reports. These could serve as evidence for an ASDE if the discrepancies in start times of totality could be resolved.

All Shall Hide

The discrepancies between the start times of totality can be resolved to yield a set of consistent observations. From the eleventh century through the twelfth, Europe had changed the meaning of the 9th hour to represent Noon.[163] Applying this fact to the A.D. 1133 reports, totality had started midday in Kerkrade, Augsburg, Heilbronn, Zwiefalten, and Salzburg. Only the report of the start time from Reichersberg remains inconsistent with the other reports. These should be considered as evidence for an ASDE. The corresponding irradiance level for A.D. 1133, according to the modified Perry-Hsu solar activity model, is 16.60%.

(9.) *A.D. 1178 September 13.* Stephenson cited a report from Baghdad indicating a solar eclipse had started in the afternoon and had retained totality until the beginning of sunset.[164] Stephenson presented another report from Vigeois, France, indicating the totality of a clear-day partial eclipse had spanned from the fifth hour to the sixth; had allowed Venus to appear; had suntanned the faces of the witnesses; and the solar disk had been darkened and re-illuminated from east to west.[165] The easterly darkening and relighting of the solar disk was contrary to Moon's path of covering and uncovering the Sun during total and partial eclipses. Figure 3.18 in *Historical Eclipses and Earth's Rotation* is a map of the fourteen separate locations of observers who had recorded the eclipse.[166] Adjustments to the path of totality failed to enclose all fourteen observer locations. Therefore, this can be treated as evidence for an ASDE. The corresponding irradiance level for A.D. 1178, according to the modified Perry-Hsu solar activity model, is 14.68%.

(10.) *A.D. 1239 June 03.* The following reports were attributed to the total solar eclipse of A.D. 1239 June 03 by Stephenson, with the exception of the account from Lucca, Italy. The report from Coimbra, Portugal,[167] indicated the Sun was black from the 6th to the 9th hour with the Moon and stars visible. According to the chronicler in Toledo, Spain,[168] the Sun was obscured and had lost its strength from the 6th to the 9th hour. Montpelliar, France,[169] reported the Sun was eclipsed from midday to the 9th hour. Florence, Italy,[170] the Sun was obscured for several hours beginning at the 6th hour and the stars appeared. Stephenson included the report from Arezzo, Italy,[171] stating the eclipse had started at the 6th hour and had endured for the time it normally takes a man to walk 250 paces. Siena, Italy,[172] indicated the Sun was obscured at the 6th hour through the 9th hour and "people lit lamps in houses and shops." According to Cesena, Italy,[173] the Sun was blackened at the 9th hour for approximately an hour with a fiery aperture at its disk. The eyewitness report from Lucca, Italy,[174] stated the eclipse of the Sun had started at 3:00 o'clock and the Podesta of Lucca had led a procession of clergy and friars through the streets to calm the frightened populace. An inscription on a pillar in Marola, Italy,[175] stated the Sun had died at the 9th hour. The report from Spit, Croatia,[176] stated a terrible eclipse of the Sun had occurred in a clear sky.

The discrepancies between the start times of totality can be resolved to yield a set of consistent observations. From the eleventh century through the twelfth, Europe had changed the meaning of the 9th hour to represent Noon.[177] The A.D. 1239 darkening had transpired during that transition and the translations of the reports stating the 9th hour or 3:00 o'clock start time should have been reworded to Noon. And, the other discrepancy, Arezzo, Italy, should be removed because its report was not dated.[178] The corresponding irradiance level for A.D. 1239, according to the modified Perry-Hsu solar activity model, is 9.05%.

(11.) *A.D. 1241 October 06*. According to Stephenson, Reichersberg, Austria, indicated the Sun was suddenly darkened after midday, for a period of four hours, and the corresponding report from Stade, Germany gave a similar start time.[179] Stephenson indicated this eclipse was reported in the Nile Delta, Egypt, to last an hour, transpiring from the middle of the 8th to the middle of the 9th hour, and many people had lit their lamps.[180] His account of the report from Split, Croatia, indicated the people had been terrorized by the darkness.[181] The reported lengths of totality from Europe and Africa can serve as evidence for an ASDE. The corresponding irradiance level for A.D. 1241, according to the modified Perry-Hsu solar activity model, is 9.06%

(12.) *A.D. 1361 May 05*. Johnson indicated Constantinople had recorded a large solar eclipse for this date.[182] The quotation made by Pang, Yau, and Chou, taken from the *Songjiang District Gazette*, stated the Sun had been suddenly darkened near sunset, its obscuration was shaped like a banana leaf, stars had become visible, totality had lasted an instant, and that it was observed in two Chinese cities that were separated by 200 km: Songjiang and Tiantai.[183] Stephenson's interpretation of the *Sung-chiang Fu-chih* was similar to Pang's with the exception that the translation of the term describing the duration of totality meant "for the duration of a meal."[184] Consuming a meal normally takes much longer than the calculated five minutes of totality. For the same date, Stephenson indicated the *Koryo-sa* had reported a total solar eclipse for Songdo, Korea – a city that was 1,000 km northeast of Sung-chiang (Songjiang).[185] The solar eclipse for that date was not able to satisfy the details about the recorded length and intensity of darkness in Sung-chiang, Tiantai, and Songdo. The path of totality that had passed between Tiantai and Sung-chiang had not been wide enough to allow reports of totality from both of those cities. The Songdo observation had been too far north of the eclipse to justify its report of totality. Therefore, these records could be considered as evidence for a mild form of ASDE – partial solar darkening event. Two months before this ASDE, there was an observation, out of Korea, of intense sunspot activity on March 16, 1361.[186] With respect to solar activity periods, this falls within the Late Medieval Maximum (A.D. 1350 – 1410).[187] The corresponding irradiance level for A.D. 1361, according to the modified Perry-Hsu solar activity model, is 6.04%.

(13.) *A.D. 1514 August 20.* Stephenson used only the two accounts from the Chinese provinces of Fu-chien and Chiang-hsi to represent the ten reports about the solar eclipse of A.D. 1514 August 20. His quotations of the two reports indicated totality had suddenly started at *wu*, the midday double-hour; had lasted for at least two hours; and had frightened birds, cattle, and people.[188] This should be considered as evidence for an ASDE because the independently reported lengths of totality had exceeded 7 minutes 31.1 seconds. With respect to solar activity, this occurred four years after the Spörer Minimum (A.D. 1400 – 1510). The corresponding irradiance level for A.D. 1514, according to the modified Perry-Hsu solar activity model, was 6.74%.

Section 2-6. Deductions

Interpreting Revelation 6:12 as a description of the Sun literally darkening in the future is not preposterous. The survey of the literature, conducted as a test of the ASDE hypothesis, identified twelve historical accounts of the Sun blackening in addition to the darkness of the crucifixion of Jesus Christ. The details of those reports could not be attributed to ecliptical, meteorological, and/or volcanic phenomena.

The findings from the tests were in good agreement with assumptions of the ASDE hypothesis. The A.D. 1361 May 05 dimming of the Sun was lighter than other accounts, but was dark enough to allow the stars to shine. The tests for ASDEs identified reports that can serve as evidence for the existence and reoccurrence of solar activity that can cause severe reductions in sunlight. The assumption that ASDEs transpire only during periods of high solar activity was not supported by the findings. Only eight of the thirteen ASDEs occurred during a solar maximum activity period. None of the ASDEs transpired during a solar minimum activity period. Most of them had irradiance levels from the modified Perry-Hsu solar activity model that exceeded the 8.02% of the crucifixion. The ASDE with the lowest irradiance level came from the detailed report of A.D. 1361 of the darkened segment of the Sun shaped like a banana leaf. Seismic disturbances were reported only for the A.D. 33 and A.D. 1133 ASDEs.

The rarity of a phenomenon has not been a strong argument against its reality. Several facts have been established by events with low occurrence rates. For example, on July 21, 2000, the collaboration of 54 scientists from Greece, Japan, Korea, and the United States announced the first direct evidence for the tau neutrino.[189] At that time, only four interactions out of the 6,600,000 triggered events had been attributed to the third class of neutrinos predicted by the Standard Model of elementary particle physics. Its occurrence rate was one out of 1.65 million. For a comparison, consider an estimated occurrence rate for ASDEs. The year of the earliest record of a solar eclipse was BC 763. Approximately 1,012,145 days would have elapsed by 2010 AD. Since thirteen of those days featured records of acute solar darkening events exceeding thirty minutes, the corresponding occurrence rate would be one out of 128,000. ASDEs may seem to be very rare, but their occurrence rate is more than twelve times greater than the corresponding rate for the tau-neutrino tests.

Other phenomena that may be attributed to the heliophysical effects associated with an ASDE were the reports of frighten birds and cattle for the A.D. 1514 ASDE and the tanned faces from the A.D. 1178 ASDE. A heightened intensity of ultraviolet (UV) light during the darkness may have caused the tanning's described in the latter. The UV intensities may have increased to offset the diminished flux of visible light. With respect to the former accounts, intensified variations in the Earth magnetic field, associated with the A.D. 1514 ASDE, may have alarmed the birds and cattle. Several studies have shown that animals can sense the magnetic field of the Earth[190] and their ability to navigate[191] has been influenced by magnetic storms. A geomagnetic storm and the sudden darkness associated with that ASDE may have been the cause of their freight. A darkened Sun is not a quiet Sun.

A feature of the ASDEs that was not included in the hypothesis was the suddenness of the onslaught of darkness. The detailed review of the darkness at the crucifixion revealed a rapid darkening feature. A GoogleBook search of Stephenson's *Historical Eclipses and Earth's Rotation* yielded ten cases out of thirteen where the words sudden or suddenly were used in the descriptions of ASDEs. Therefore, it is reasonable to expect a rapid drop in sunlight with the fulfillment of Revelation 6:12.

The scientific community is likely to dismiss the evidence for ASDEs until one occurs. Consider the history of the acceptance of meteorites. Luke's description of one of the objects of worship by the Ephesians became one of the best known biblical accounts about a meteorite (Acts 19:35). During the 18th century, the French academy adopted the policy of dismissing records of meteor sightings and meteorites. They had motivated other institutions such as museums to implement similar practices of rejection and ridicule. It took the meteor shower of 1803 at L'Aigle, near Paris, to convince the Paris academy to recognize the reality of falling meteors.[192] The study of meteor astronomy started after the spectacular Leonid meteor shower of November 13, 1833.[193]

The scientific community will become more receptive to examining and incorporating ASDEs as an additional class of solar activity whenever they detect similar behavior in Sun-like stars. The NASA Kepler Mission has the capability to record, with statistical significance, an ACDE among the 100,000 stars it is monitoring. Data retention protocols should be created to capture such events. Repeated reductions in brightness are the primary pattern the Mission was designed to document. Current Mission priorities could result in either the misclassification of an ACDE as component failure or data deletion.

The historical trustworthiness of the New Testament justified its use as a source of physical facts about the darkness at the crucifixion. Evidence for the reality of ASDEs suggests the need to reexamine at least Joel 2:31; Acts 2:20; and Revelation 6:12-17.

Taylor A. Cisco, Jr.

Prospective Acute Solar Darkening Events (ASDEs)

Date	Darkness Start	Stop	Length	Listed as an Eclipse	Primary Report Locations	Solar Activity Epochs
BC 645 Apr 03	...	...	...	NO	Ch'ü-fu, China, and Thasos or Paros Island, Aegean Sea	4th Grand Maximum in the SN-S Series [BC 640 - 550]
BC 592 May 15	...	...	...	NO	Ch'ü-fu, China, and (2) Asia Minor	
BC 480 Early Spring	...	...	...	NO	Asia Minor and Thebes.	3rd Grand Maximum in the SN-S Series [BC 490 - 420]
BC 34 Aug 23	...	...	...	NO	(2) Ch'ang-an, China.	...
AD 33 Apr 03	12 P	3 P	3 hrs	NO	(4) Jerusalem, Israel; and Nicaea, Bithynia.	Roman Maximum [BC 20 - AD 80]
AD 879 Apr 25	3 P	...	1 hrs	NO	(2) Ch'ang-an, China, and England	...
AD 977 Dec 13	...	...	...	NO	(2) Pien, China.	...
AD 1133 Aug 02	12 P		<1 hrs	YES	England; Kerkrade, Netherlands; Augsburg, Heilsbronn and Zwiefalten, Germany; and Reichersberg and Salzburg, Austria	Medieval Maximum [AD 1120- 1280]
AD 1178 Sep 13	11 A	12 P	1 hrs	YES	Vigeois, France, and Baghdad.	
AD 1239 Jun 03	12 P	3 P	3 hrs	YES	Coimbra, Portugal; Toledo, Spain; Montpellier, France; Lucca, Florence, Siena, Cesena, and Marola, Italy; and Split, Croatia.	
AD 1241 Oct 06	>12 P	3 P	<4 hrs	YES	Split, Croatia; Reichersberg, Austria; Stade, Germany; and Cairo, Egypt.	
AD 1361 May 5			Length of a Meal	YES	Constantinople; Sung-chiang and Tiantai, China; and Songdo, Korea.	Late Medieval Maximum [AD 1350 - 1410]
AD 1514 Aug 20	wu	yu	> 2hrs	YES	Fu-chien and Chiang-hsi provinces, China. Over ten separate descriptions.	

Section 2-7. Black as Sackcloth of Hair

Can the blackened Sun possess a textured image? It may be invisible to the naked eye, but its image through a telescope may resemble a woven surface like sackcloth. The answer to this question may stem from current observations of the sunspots and the surface of the Sun through special light filters. Sunspots appear black in white light. Sunspots are not literally black. Their image appears to be black because they are cooler than the surrounding surface (photosphere) of the Sun. The photosphere is the region of the Sun that produces visible light and functions at a temperature of approximately 6,000 K. The temperature of sunspots is around 4,200 K.

During the seventeenth century, early refracting telescopes, including those developed by Galileo Galilei, allowed astronomers to observe the penumbra and umbra of sunspots. The umbra is the darkest region of a sunspot and its penumbra is the surrounding gray zone. Penumbrae are between three to five times larger than the area of the umbrae.[194]

The light bridges through the darkest areas of sunspots may be a clue to the future sackcloth texture of the blacken Sun. Considerable information about the Sun has been obtained by viewing it through Hα (Hydrogen alpha) light filters. Hα images of the photosphere depict a grainy surface. Hα pictures of penumbrae reveal granules that are lying on their sides and filaments radiating from the umbrae. Similar pictures of umbrae exhibit umbral dots and light bridges. Large sunspots feature complex arrangements of filaments that extend across the umbrae. The sackcloth image may be caused by patterns of light bridges to resemble a woven texture.

Another clue may stem from the structure of the photosphere. "One can think of the magnetic field beneath the photosphere of the sun as a giant ball of yarn."[195] The Revelation 6:12 ASDE may cause the string-like nature of the fields beneath to photosphere to shift to a woven pattern. Hughes, Paczuski, Dendy, Helander, and McClements (2002) have proposed a magnetic carpet as a model of the photosphere.[196] It treats the stability of the random distribution of magnetic loops as a product of self-organized criticality. Such a model allows the crisscross arrangement of magnetic flux tubes that may be the source of the sackcloth of hair image of the Sixth Seal ASDE. The stability of those distributions may imply an unusually large period of darkness. Revelation 6:15 indicated the entire population of the world flees to shelters. The darkness of the blackened Sun may last long enough for all people to successfully reach and hide in dens and caves. How long would it take the world to achieve such an accomplishment?

Chapter 3
The Moon Becomes As Blood

And I beheld when he had opened the sixth seal, and, lo, there was a great earthquake; and the sun became black as sackcloth of hair, **and the moon became as blood;** (Revelation 6:12d)

Section 3-1. The Black Sun With Red Moon Sign

The Moon as blood is the distinctive feature of this prophecy. It is not the most dramatic feature, but it sets this scripture apart from other solar blackout predictions. According to Isaiah 13:9-10, an additional solar darkening will be accompanied with losses in both lunar and star light during the Day of Wrath. The loss of light from both the Sun and Moon denotes Christ's coming (see Matt. 24:29-30 and Mark 13:24). But, the combination of the Moon as blood and a black Sun is limited to heralding the great Day of Wrath. During the Day of Pentecost, Peter included Joel 2:31 with his explanation for the behavior of the Holy Ghost filled believers. Joel's explicit list of pre-wrath signs ended with the solar blackout and the blood colored Moon. Correspondingly, other solar blackout scriptures can not supplement Revelation 6:12 if the Moon is not compared with the appearance of blood.

Transmutations have a prominent role in the *Holy Bible*. In the beginning, God created water out of nothing and then transmuted the water into the Earth (Genesis 1:2, 9; 2 Peter 3:5). Two of the three signs to the Jews and the Pharaoh that identified Moses as a prophet of Jehovah were transmutations of the rod of Aaron into a serpent (**Exodus**

4:1-5; 7:10-13) and turning clear water into blood (Exodus 7:19-22). Changing water into wine at a wedding feast was the first recorded miracle performed by Jesus Christ (John 2:1-11). But, the "moon became as blood" clause does not explicitly denote a transmutation.

Several factors justify treating Revelation 6:12 as a description of color, not a change in molecular structure. The article "as" in the phrase indicates the Moon resembled blood. If the phrase had been the moon became like blood, the scripture could be interpreted as another transmutation. Other translations of the *Holy Bible* (e.g., NIV, LB, and JB) clearly indicate the Moon acquired the color of blood. With respect to the physical science aspect of the description, a moon-size ball of blood would be hard to see with the absence of visible light from the darkened Sun. And, the capabilities of the sensors left on the Moon during the Apollo Program may not be able to verify a transmutation. Since the pre-wrath model has been developed from physical science aspects, it assumes the Moon emits dark red light.

The Moon has acquired a red color on numerous occasions. The smoke from fires and volcanic eruptions had filtered moonlight to red and dark orange hues. During lunar eclipses, the Moon passes through the penumbra of Earth's shadow caused by the Sun. Light in the penumbra region has a reddish hue because it has been filtered by the Earth's atmosphere. Subsequently, the Moon reddens as it passes trough the penumbra. The Moon becomes dark when it reaches the umbra of the shadow. But, the Earth's shadow will cease when the Sun temporally stops radiating in visible light. Since an eclipse can not occur in the absence of sunlight, the reddening predicted by Revelation 6:12 can not be caused by a lunar eclipse. Also, the examples given above within this paragraph use light shining through a filtering medium.

Section 3-2. Lunar Luminescence

How can the Moon glow red in the absence of sunlight? We were taught in school moonlight was reflected sunlight. Revelation 6:12 states lunar reddening follows a solar darkening. This seems to be a contradiction. An answer to this question can be found in the following chronology of attempts to explain observations of peculiar phenomena on the Moon.

A good hint of the answer emerged on the night of April 19, 1787.

All Shall Hide

An aurora borealis (northern lights) rippled above Padua, Italy.[197] Aurora activity that far south from the Arctic Circle was very rare. Padua's display happened a few days before the sunspot number had peaked in May 1787. During the night of April 19th the famous British astronomer, Sir William Herschel, noticed three red glowing spots on the dark part of the Moon.[198] He informed King George III and other astronomers of his observations. Sir William attributed the phenomena to erupting volcanoes and perceived the luminosity of the brightest of the three as greater than the brightness of a comet that had been discovered on April tenth. Those areas emitted a stronger glow on April 20th and Herschel determined the diameter of the brightest spot to be three miles. A comparison he made between his observations of April 20, 1787, and May 4, 1783, yielded a greater luminosity area for April and brighter glow for May. The three sets of observations occurred on the dark side of the Moon and the red glowing spots continued to occur for weeks in April 1787.[199] The red glowing regions appeared inside and near the Crater Aristarchus. They appeared again in 1788. Sir William Herschel observed a red brightening, in the absence of sunlight, in three lunar areas.

Various types of physical phenomena had caused the red glows and other observations. Only one family of processes could cause the entire lunar surface to glow without sunlight.

During the following centuries, reports appeared about different colored glows, flashes of light, and obscurations of lunar features. These came from a diverse variety of observers. Since the phenomena were brief and unpredictable, they acquired the classification of transient lunar phenomena (TLP). The absence of photographic evidence rendered the TLP reports to discreditable merit. Many astronomers tried to attribute TLP reports to misunderstood atmospheric phenomena. Over 350 reports of TLP were made from the year of Sir Herschel's observations to 1920 – the year when the search for relationships between solar activity and lunar eclipses had begun.

In 1920, French astronomer Andre-Louis Danjon published a correlation between the sunspot cycle and the residual brightness of the Moon during a total lunar eclipse.[200] Residual brightness was greatest when the number of sunspots in the lower solar latitudes was at maximum and decreased sharply when the sunspot number diminished.

His deduction stemmed from analyses of three and a half centuries of lunar eclipse data. In 1933, Rougier published a correlation between the one-day lag in variations in the excess brightness of the sunlit side of the Moon and the solar constant.[201] In 1947, Czech astronomer, Frantisek Link, proposed lunar luminescence, driven by energetic particles flowing from sunspots, as the cause of the variations deduced by Danjon and Rougier.[202] The corpuscles were slower than light and followed paths dictated by magnetic fields. Link realized such particles could travel a curved path around the Earth to hit the Moon during a lunar eclipse. Visible light can not follow such paths. Luminescence excited by energetic particles from the Sun was one of the keys for explaining a future lunar reddening in the absence of sunlight

Several mechanisms can cause light. The process of heating an object to the point of forcing it to emit light is incandescence. With increased heat, an object can be caused to glow red, then yellow, white, and, with more heat, blue. The color and intensity of light produced by incandescence is described by the Stefan-Boltzman temperature formula. Sir William Herschel had assumed the glowing red areas on the Moon had been caused by incandescent emissions from hot lava.

The mechanism Link had proposed as the cause of excessive lunar brightness, luminescence, was not driven by heat. Luminescence occurs when objects are stimulated to emit light without generating heat. Such objects are known as phosphors. Stimulated phosphors emit light in narrow bands of the spectrum. That is, phosphors emit a particular color, not a spectrum of various colors.

Many different modes of excitation cause objects to emit light without heat. Television sets of the 20th century used cathodoluminescence (CL) to display pictures. Electrons bombarded the phosphors in the TV screens to cause pictures by the CL process. Photoluminescence occurs when excitation by one color causes a substance to emit a different color. Several parties have used black lights (ultraviolet light) to cause clothing and make-up to visibly radiate in entertaining hues. The light emitting diodes (LED) in calculators use electroluminescence: stimulated emission of light by alternating electrical currents. Chemical reactions that produce light fall in the category of chemoluminescence. Emergency sticks, that are shaken to produce light, are examples of chemoluminescence. Bioluminescence is caused by biochemical reactions and triboluminescence is driven by mechanical disruption. Producing light by penetrating a surface with energetic ions is known as ion beam induced luminescence, or, more simply, ionoluminescence (IL). Lund University, Sweden, and MARC Melbourne, Australia, developed IL analytical methods during the 1990's. Phosphorescence occurs when after-glows persist longer than 10^{-8} seconds after excitation. Examples of phosphorescence are toys, stickers, and watches that continue to glow for minutes or hours after excitation. Luminescent after-glows shorter than 10^{-8} seconds are classified as fluorescent.

Astronomers developed tests to directly evince luminescence as the cause of excess brightening. In 1950, Link was the first one to propose the "method of line depths" for detecting luminescence in various colors of moonlight.[203] The ratio of lengths of selected Fraunhofer spectral lines to the adjacent continuum from sunlight were compared with the corresponding ratio of the same line to continuum in moonlight. If luminescence were not present, the ratios would be identical.

During the latter half of the 1950's, Nikolai Kozyrev, Pulkovo Observatory, former USSR, and Jean Dubois, Bordeaux Observatory, France, published the earliest analyses of photographic spectrographs. Their application of the "method of line depths" yielded positive evidence with inconclusive accuracy.[204]

Although many researchers have cited Kozyrev's reports, his works in the H and K ultraviolet spectral lines in ionized Calcium (Ca II) have received severe criticism.[205] They reported spectral enhancement of those lines for the region of craters Aristarchus and Herodotus (10/04/55), red spectral enhancements for the craters Alphonsus (11/03/58 and 10/23/59) and Aristarchus (11/26/61, 11/28/61, and 12/03/61).[206] He attributed some of the phenomena to various forms of volcanic outgassing. For example, A. A. Kalinyak's spectral analyses of Nikolai's spectral photographs taken of the Alphonsus crater reddening yielded a release of one million cubic meters of gas containing molecular carbon.[207] Nikolai Kozyrev's publications and press conferences caused intense interest and alarm in the astronomical community. Subsequently, extensive federal funds were appropriated to American scientists studying the solar system.[208]

Gehrels, Coffeen and Owings also reported the reality of luminescent phenomena.[209] Measurements of surface brightness of certain regions of the Moon were both 27% higher and reddish in November 18, 1956 than in December 30, 1963.

Dubois reported indications of luminescence for about half of the ninety areas of the Moon that he had examined in the D, E, F, G, and Hα visible lines of the Fraunhofer spectra.[210] He found the portion of moonlight intensity that could be attributed to luminescence ranged from 3 to 25%. Their fluctuations at various sites followed the time variations in different types of solar emissions. The Dubois and Kozyrev reports were the first to qualitatively evince the production of moonlight without the use of sunlight. Sunlight served to determine relative intensities of luminescent light.

Incontrovertible evidence of lunar luminescence was secured after the development of spectrometers that could accurately measure intensity increases as small as 1%. The production of a photo-electric spectrometer for the investigation of lunar luminescence by the "method of line depths" was completed by John F. Grainger and James Ring, University of Manchester, England.[211] The measurements from their device, attached to the 50-inch telescope of the University of Padua's observatory at Asiago, Italy, quantitatively confirmed the existence of lunar luminescence beyond any shadow of a doubt. Their analyses in 1961 incorporated the H line of Ca II (3970 Å) to detect luminescent emissions from the crater Aristarchus (May 30-31); the bright ray crossing Mare Senitatis through the crater Bessel (June 27-28); and East of the crater Plato.[206] Intensities varied with time and reached $10 \pm 1\%$, but could not be correlated with solar events.[212] Their research was supported by the United States Air Force under Contract AF61 (052)-378.

Robert Wildey and Howard Pohn photoelectrically investigated twenty-five lunar features on three nights of April, one in May, one in June, three in July, one in August, three in September, and two in December 1962 and one in January 1963 with the 60-inch telescope of the Mount Wilson Observatory.[213] They did not attribute their recorded photometric discrepancies to lunar luminescence. Eighteen years later, rigorous analyses of their work by Winifred Sawtell Cameron, National Space Science Data Center, NASA's Goddard Space Flight Center, proved each of the fifteen nights of observation had significant color and emission anomalies.[214]

Confirmations for Grainger's and Ring's observations came from independent observations, on September 16, 1962, when Hyron Spinrad, Dominion Astrophysical Observatory, detected luminescence using the H and K lines of Ca II.[215] The luminescent phenomena appeared to be widespread across the Moon with 13% of the intensity, at the wavelength of 3950 Å. Spinrad was not able to identify an energy source for the phenomena. A moderate magnetic storm that had erupted September 12th was the sole form of detectable solar activity that he thought may have caused the luminescence.

Incontrovertible evidence for lunar luminescence in the prominent ultraviolet lines of the Sun's spectra had been established. The accurate measurements by Grainger, Ring, and Spinrad were in agreement with Kozyrev's findings. Numerous reports and accurate evidence of luminescence in the visible and near infrared lines of the spectrum were acquired the following year. Photographs of features glowing red were obtained through non-spectrographic methods.

On March 11, 1963, Wildey used a high-resolution photoelectric spectrum scanner and spectrograph camera of the 100-inch telescope to test eight lunar features for luminescence with the "method of line depths".[216] He treated the results as negative since it yielded 2% excess in the brightness of the H spectral line of Ca II.

Scarfe, University of Cambridge, applied the "method of line depth" analyses to eighteen lunar sites, through ten visible spectral lines, during the autumn and winter months, from 1963 to 1964.[217] A strong luminescent glow (30% intensity) was detected during the night of October 5, 1963, for a sector of the Moon's surface between Mare Serenitatis and Aristarchus, in the near green spectral line of Fe I (5450 Å). Moderate luminescence (13±5% intensity) was recorded for Aristarchus during November 2, 1963 and December 4, 1963 in the Fe I line.

He was not able to associate the luminescence with any solar activity. Scarfe indicated the Sun was relatively quiet from October 1963 through February 1964; there were no reports of solar flares for the preceding days of October second and third; and there were reports of two flares erupting approximately eighteen minutes before the luminescence events. Scarfe's measurements were accurate, but the explanation was a mystery to him. He had indicated the pair of flares occurred between 23.02 hr and 23.41 hr UT and the luminescence followed between 23.35 hr and 00.45 hr UT. The corresponding transit times from those flares to luminescent glow yielded speeds from 129,861 km/s to 39,957 km/s. For protons, those speeds corresponded to kinetic energies ranging from 101.7 MeV to 8.4 MeV, respectively. Since most astronomers had expected solar wind speeds to approximate 440 km/s, Scarfe did not attempt to connect the two events. It is now known that solar energetic proton (SEP) events are real, rare, and erupt with kinetic energies beyond 100 MeV. Scarfe's observations were probably caused by SEPs.

It should be noted that the Fall of 1963 corresponded to the minimum phase of a sunspot cycle. Astronomers had expected calm solar activity. But, a large sunspot group appeared and caused spectacular Auroras during the 21st, 22nd, and 24th of September.[218] The intensity of the aurora and the extent of its southern limits indicated the strength of the associated geomagnetic activity. A copy of Dr. Carl A. Gartlein's map in *Sky and Telescope* featured a southern limit corresponding to Kp = 9 for the auroral area.[219] Since the range of the K-index scale is from 0 to 9, the geomagnetic activity associated with that aurora and the green luminescent glows were very strong.

A different type of proof for the Moon's ability to luminesce began to emerge during the same period. In 1960, Geake, Lipson, and Lumb had begun to assemble equipment to simulate lunar luminescence in a physics laboratory at the Manchester College of Science and Technology.[220] Support for their project came through Contract AF61 (052)-379 between the Department of Astronomy, University of Manchester and the Geophysics Research Directorate, Air Research and Development Command, U. S. Air Force, European office in Brussels. By 1963, Derham and Geake had discovered certain types of stony meteorites would luminesce in red.[221] They had tested a variety of materials that had been assumed to exist on the Moon. Most of the samples generated negligible emissions under bombardment of 40 keV protons. But, three of the samples had strong luminescent emissions in the red. Those three were meteorites that had fell to Bishopville, Bustee, and Khor Temiki in 1843, 1852, and 1932, respectively. They belonged to the mineral classification of enstatite achondrites. Achondrites are stony meteorites, having the texture of some terrestrial rocks that do not contain little seed-shaped chondules. Enstatites are inosilicates with the crystalline form of orthorhombic prisms. The red spectrum from the proton excited luminescence had a width of 900 Å around a peak that reached maximum at approximately 6,700 Å. Three meteorites luminesced slightly in the blue with a slight peak occurring near 4,000 Å. The Bustee sample had the highest emission level. Subsequently, their rough estimate for phosphor efficiency was 20%. Under the assumption that various areas of the Moon would be covered with enstatite achondrite dust, Derham and Geake suggested searching for lunar luminescence in the Hα spectral line (6562 Å). Subsequently, a team of astronomers, Kopal and Rackham, engaged the task of implementing their suggestion.

On October 29, 1963, two Aeronautical Chart and Information Center (ACIC) cartographers, working for the U. S. Air Force lunar mapping program, recorded a set of phenomena that would escalate research into TLP's. The two ACIC cartographers, James A. Greenacre and Edward Barr, at the Lowell Observatory, Flagstaff, Arizona, observed very bright red, orange, and pink color phenomena on the southwest side of Cobra Head; a hill southeast of Schroeter's Valley;[222] and the southwest interior rim of the crater Aristarchus.[223] They viewed the brightenings through the 24-inch refractor telescope. At 6:50 PM, during full solar illumination of the Moon, Greenacre noticed the reddish orange glows at Cobra Head and near Schroeter's Valley. After he had called Edward Barr to witness the phenomena, they removed the Wratten 15 (deep yellow) filter from the telescope. The color remained; the images were much brighter, and sparkled. Those features convinced Greenacre of the reality of the event. He had been among the skeptics of TLP reports. Greenacre began to scan neighboring features for similar intense lights and, at 6:55 PM, discovered a pink radiance in the interior southwest rim of Aristarchus. All of the features began to pale gradually at 7:05 PM. The first two glowing features had dimmed to their normal colors by 7:10 PM and Aristarchus five minutes later. They did not have enough time to cease their observations to expend minutes to refocus the refractor's 70-mm camera for a sharp image. Greenacre then recorded his detailed observations on tape after the phenomena had ceased.

This event sparked a change in attitude towards TLP reports. According to the rocket historian Willy Ley: "The first reaction in professional circles was, naturally, surprise, and hard on the heels of the surprise there followed an apologetic attitude, the apologies being directed at a long-dead great astronomer, Sir William Herschel."[224] Many lunar researchers, during the 1960's, believed the Moon had been formed as a cold body through meteor impacts.[225] Winifred Sawtell Cameron stated: "This and their November observations started the modern interest in observing the Moon."[226] The credibility of their findings stemmed from Greenacre's exemplary reputation as an impeccable cartographer. It is interesting to note that this monumental change in attitude by professional astronomers had been caused by the reputation of a researcher and not by the acquisition of photographic evidence.

Section 3-3. Wide Area Reddening

A few days later, at the Observatoire du Pic-du-Midi in the French Pyrenees, photographic discoveries were made that supported Derham's and Geake's hypothesis. Zdenek Kopal and Thomas Rackham,[227] Czech-American and British astronomers, respectively, made the first photographs of wide area lunar luminescence on the night of November 1-2, 1963. Their achievement fortified arguments to use the Moon as a research tool for gathering information about the energetic particles in space. With respect to biblical prophecy, Kopal and Rackham obtained photographic evidence of a mechanism that can redden the Moon during the absence of sunlight.

Kopal's enthusiastic drive towards placing men on the Moon won copious funds from the U. S. Air Force.[228] The lunar luminescence research was a component of Kopal's lunar mapping program that had commenced back in 1958. He had acquired funding for that mapping program three years before President John F. Kennedy had declared the commitment to place a man on the Moon. Acquisition of accurate, comprehensive three-dimensional data for maps that may be needed for manned lunar landings was the primary goal of Kopal's program. Its funding came through Contracts AF61 (052)-168, 380, and 496, between the Department of Astronomy, University of Manchester and the Geophysics Research Directorate, Air Research and Development Command, U. S. Air Force, European office in Brussels.[229] Specific support for their tests for luminescence came through Contracts AF61 (052)-378 and 400. The family of AF61 (052) contracts included the appropriations that had been accorded for the construction of Derham's and Geake's instruments. Kopal's enthusiasm and prayers had successfully diverted resources towards the development of projects that would discover numerous relationships between interplanetary plasma and the Moon through luminescence. Kozyrez's studies had convinced Kopal that densities and kinetic energies of the solar protons were more than sufficient to cause lunar luminescence.[230]

They had modified their equipment to test the luminescence hypothesis by Derham and Geake. Red and green narrow-passband interference filters centered at 6,725 Å and 5,450 Å, respectively were used on the 24-inch refractor telescope to photograph the northwestern part of Oceanus Procellarum. Near infrared light from the enstatite achondrite samples had a broad band half width of 900 Å and a peak at 6,700 Å in Derham's and Geake's examinations. Their filter, centered at 6,725 Å, with a half width of 45 Å, could capture similar emissions. Pairs of photographs, taken in rapid succession, were to serve as tests for red brightening. If luminescence did not occur, the intensity of light from pairs in different colors would be the same for identical lunar features. Pictures taken through the red filter would yield brighter images, if red luminescence happened.

Eight pairs of photographic plates were taken from 22:35 UT on November 1, 1963 through 00:35 on November second of the Oceanus Procellarum plain.[231] That region of the Moon contained the Aristarchus, Kepler, and Copernicus craters. Their discovery of wide area, red brightening was made after the plates were developed. The first and second pairs manifested brightening through the red filter; pairs three and four yielded no red enhancement; and the last four pairs had images of the second brightening in red. Analyses of the plates with a microdensitometer yielded an 86 ± 3 percent enhancement of the Kepler region! The images had captured luminescence for an area of 60,000 square kilometers. A search for possible solar causes identified two eruptions of moderate flares. The spacing between the flare events corresponded to the spacing between the photographed brightenings. They occurred 8 ½ hours before the brightening. Unfortunately, the required energy and density of protons to cause that level of luminescence seemed to be much too high for those flares. Kopal and Rackham shared the predicament of other astronomers who had carefully recorded unexplainable phenomena.

The inability to provide an excitation mechanism for luminescence caused some to attribute the observations to experimental error. For example, Ney, Woolp, and Collins suggested Kopal's and Rackham's observations were inconclusive because some of their type of interference filters lacked transmission uniformity and the images of luminescence were captured near the edge of the photographic plates.[232] Astronomers continued to carefully record the inexplicable events.

Edward Barr[233] discovered another anomalous illumination of Aristarchus on November 27, 1963. This event lasted for 75 minutes, covered a larger area (12 miles by 1½ miles) of the crater rim, and had a ruby flush color. He summoned James Greenacre, the Director of the Lowell Observatory, Dr. John S. Hall, and Fred Dungan to witness the red glows. Dr. Hall called Peter A. Boyce of the Perkins Observatory to look at the phenomena near Aristarchus with the observatory's 69-inch telescope.[234] Hall was careful not to give Boyce the location of glowing area. Boyce called back the precise coordinates of the glowing region. They confirmed Greenacre's measurements. Attempts were made to photograph the event with black-and-white film. They failed to appear on film and densitometers tests were negative. Another observer tried to see the phenomena with a 12-inch telescope. His attempt failed. The most thorough report of Greenacre's and Barr's observations was published in May 1964 for the funding source of their works, the U. S. Air Force. It contained their descriptions of additional color changes that followed three hours after the red glows of October and November. In both events, a region of violet and purple blue hues appeared and spread along the rim of Aristarchus.

Many observers witnessed anomalous brightening phenomena during the lunar eclipse of December 1963. On the evening before the lunar eclipse, December 29th, Y. Yamada and eight observers saw an area covering the southern region of Aristarchus glow pink and spread towards Herodotus.[235] They had assembled at the Rakurakuen Planetarium, Hiroshima, Japan, to practice for the eclipse that was to occur on the following night. During the darkest portion of the Earth's shadow of the total lunar eclipse of December 30th, the northern part of the Moon acquired a reddish glow.[236] Observers in Boulder, Colorado; Baltimore, Maryland; Holden Massachusetts; Bremerton, Washington; and Buffalo, New York reported it. A contrary report came from the Country Observatory of the State University of Iowa. Satoshi Matsushima and John R. Zink did not detect the reddening phenomena.[237] Their photometric tests with the 12-inch Cassegrain reflector for red enhancement were negative for Mare Crisium. If the southern extent of the observed reddish region had stopped at a longitude of 25°, Mare Crisium would not have been brightened.

Derham, Geake, and Walker (1964) published additional observations and refinements to their luminescence assessments of the enstatite achondrite samples.[238] The meteoritic samples luminesced in the red when bombarded with 2 MeV and 4 MeV protons; 5 keV to 40 keV electrons; or X-rays. Their previous study had overlooked the rapid drop in phosphor efficiency that was caused by damage to the sample by proton bombardment. Heating the samples restored their efficiency. Unfortunately, their value for phosphor efficiency had to be revised down to only one percent. That correction worsened attempts to incorporate the ejection of solar protons from the flares with the explanation for Kopal's and Rackham's observations. It spurned the creation of many theories for lunar brightenings and impaired luminescence arguments for decades. For example, Chanin *et al* provided a comprehensive proof for thermoluminescence under the assumption of insufficient solar flux.[239] Subsequent tests for thermoluminescence that were performed a year later were negative.[240]

But, astronomers continued to record the rare incidents of lunar surface enhancements. The majority of them briefly flared in localized regions such as craters, portions of maria, and/or sections of the highlands. Since the physical interpretation of Revelation 6:12-17 is the principle focus of this work, the remainder of this chapter will be limited to reports of wide area luminescent phenomena and various theoretical explanations.

The possibility of luminescence was examined as a possible explanation for anomalous variations in brightness of the entire lunar disk that had been recorded from 1964 through 1965. The Le Houga Observatory in southern France supported the photoelectric photometry program that had been directed by the Harvard College Observatory.

Le Houga did the UBV measurements. Harvard did measurements through interference filters in nine narrow spectral bands. Analyses by Adair and Irvine of the observations made from July 30, 1964 through November 7, 1965 yielded anomalous red enhancements in spectral reflectivity that could be attributed to luminescence.[241] They, like many other scientists, acknowledged the problem of insufficient solar wind energy. Lane and Irvine cited a hypothesis by A. G. W. Cameron and Speiser as a reasonable explanation. A. G. W. Cameron and Speiser were among the first to propose proton acceleration by the magnetic field of the Earth. Particular fluctuations in the geomagnetoshpere may have accelerated and focused segments of the solar wind to cause lunar luminescence. Additional research and analyses would have to be done to convert their hypothesis into a theory.

During the December 19, 1964 lunar eclipse, Sanduleak and Stock confirmed the existence of luminescence.[242] Their photometric tests at the 16-inch reflector at the Cerro Tololo Inter-American Observatory, La Serena, Chile, through a yellow filter yielded a 2.3% intensity enhancement of the entire area covering Mare Nubium.

In October 1966, Satoshi Matsushima presented a correlation between the brightness of the eclipsed moon and the geometric planetary index, K_p.[243] He found there was no correlation between the brightness of the eclipsed Moon and sunspot activity. His analyses were based on the works of Kozyrev, Spinrad, Greenacre, and Kopal and Rackham. Matsushima's analyses found that the speeds for solar wind plasma were highest for the combination of red luminescence and large Kp values. Matsushima's discussion emphasized the possible acceleration and focusing effects of the magnetosphere on the solar wind for generating lunar luminescence.

A study incorporating over two centuries of data was produced Barabara M. Middlehurst, Lunar and Planetary Laboratory, University of Arizona.[244] Her comparisons yielded no correlation between lunar events and sunspot numbers. Middlehurst's report covered over 200 TLP events from January 1749 to June 1964. Events reported by Kopal and Rackham were treated as a separate form of TLP and were excluded from her study. Her discussion favored internal stresses that were induced by Earth's gravity as the cause for radiant outgassing from the Moon's crust. One year before man landed on the Moon, NASA published Barbara's collection of more than 570 TLP sightings from over 300 observers.[245] Her collection covered various forms of transient lunar reports that could be examined for statistical trends. A general examination of her tables yields a set of areas where TLP events have frequently occurred. This fact was very strong evidence against the argument TLP's were the product of observer error.

Dunlap, et al, Corralitos Observatory, Organ Pass, New Mexico, recorded 30% enhancement for ultraviolet of the whole Moon for both nights of April 21 and 22, 1967.[246] Other astronomers made TLP reports of red and orange brightenings of Aristarchus, Schroeter's Valley, and Herodotus for those two evenings.

During the period of the manned lunar landings (July 1969 through December 1972), 840 pounds of lunar rocks were collected. Initially, the analyses of those samples did not solve the problem of deficient proton energy densities to drive luminescence in the red spectral range. But, by 1971, most astronomers had accepted the reality of lunar transient phenomena.[247] Additional tables and catalogs formed a preponderance of evidence for the reality of transient lunar phenomena.

Two Japanese astronomers recorded the beginning of a series of wide area enhancements that gave credence to Kopal's and Rackham's discoveries. During photographic photometric and polarimetric observations, Naosuke Sekiguchi[248] and Matsumoto, Tokyo Astronomical Observatory, Mitaka, Tokyo, observed an enhanced brightening of the whole Moon on March 26, 1970. Sekiguchi indicated the eruption of a large solar flare, 29 hours before the enhancement, may have been the cause of the luminescent phenomena.

Continued observations extended to three colors and standardized by the reference stars selected from the *Arizona-Tonantzintla Catalogue*,[249] enabled Sekiguchi to measure absolute variations of luminosity over the entire lunar surface. The additional filters revealed that wide area enhancements were brighter in red than in the yellow and blue colors (Sekiguchi, 1977).[250] Sekiguchi's observations from 1970 through 1975 yielded fluctuating enhancements of the entire Moon, primarily in red, that would last for a couple of consecutive days. The luminosity enhancements followed the eruptions of solar disturbances.

Sekiguchi cataloged his studies in the *Tokyo Astronomical Bulletin*.[251] His observations venerated those of Zdenek Kopal and Thomas

Rackham. Sekiguchi (1986) published additional sets of observations into the nineteen eighties.[252]

NASA's Goddard Space Flight Center published a very comprehensive list of TLP reports by Dr. Winifred Sawtell Cameron.[253] She had analyzed and compiled 1,468 reports of lunar transient phenomena that had been recorded from November 10, 557 to May 28, 1977. Some of the ancillary data that she had included to facilitate research on the phenomena were conversions of the (1.) dates and times of observations into Universal Time; (2.) selenographic coordinates of the center of the features or areas effected; (3.) the orbital location and phase of the Moon during the event; and (4.) the magnetic indexes (Kp) for the dates of the events. These and other information she had included with the observation allowed researchers to test hypotheses based on tidal forces; solar wind focusing by the geomagnetic field; and other internally and externally driven mechanisms. An interesting fact that can be extracted from her work is that the number of reports for reddenings was slightly more than twice the number of blue colored phenomena.

Approximately two hundred amateur astronomers from the Federal Republic of Germany, the German Democratic Republic, Switzerland, and Austria made 907 observations during the years 1971 and 1972 that yielded 45 lunar transient phenomena events.[254] The initial results were published during 1972 and 1973 and the total compilation of positive LTP reports was published in 1984.

Section 3-4. Analyses of Lunar Rocks

Examinations of samples retrieved from the Moon by the Apollo astronauts evinced the existence of former bursts of particles with very high energies and intensities. Analyses of lunar rock 74275 indicated some SEP's had deeply penetrated lunar surface material and carried enough energy to induced nuclear reactions.[255] Similar findings were generated by analyses of lunar rock 68815.[256] The nuclides that had been produced by SEP collisions with the lunar rocks enabled scientist to estimate the radiation fluxes during the last ten million years.[257] Lunar rock studies generated findings that could evince the existence of SEPs

All Shall Hide

with energies and fluences high enough to cause wide area red glows of the Moon in terms of luminescence.

Studies near the end of the 20th century indicated the Sun was capable of producing high energy cosmic rays. Ng and Reames developed a model that explained the acceleration of SEP to energies of 1 MeV by CME shock waves.[258] Reames' research[259] had proven coronal mass ejection (CME) shock waves, not flares, were the principle mechanism for accelerating SEP's to energies between 10 MeV and 1 GeV. One of the facts he cited in support of that argument was that flares last only a few hours and large solar energetic proton events continue for days. Further studies had been conducted to understand the correlation between shock waves, CME, and SEP.[260] The shock waves have been able to produce the fluences and energies necessary to explain the wide area lunar luminescence phenomena.

Earlier arguments against lunar luminescence were unaware of the energies and fluences of SEP that the Sun has produced. Those "gusts" in the solar wind were capable of causing wide areas of the Moon to glow red.

It is reasonable to attribute the red color of the Moon during the prophesied solar blackout to luminescence induced by solar proton bombardment. The increased output of energetic protons may be caused by the magnetic storms that transform the entire solar image into the appearance of a black globe of sackcloth of hair. A portion of those particles will be focused into concentrated streams by turbulence in the magnetic field of the Earth. Those concentrated streams of energetic particles hit the Moon to cause it to coldly glow red light. The relative uniformity of the microstructure of the Moon's surface[261] and the intensity of the solar protons would be strong enough to obscure lunar features. Subsequently, its appearance shall resemble a pool of blood. Intense luminescence would not produce shadows on the lunar surface.

The intensity of the red lunar image shall be a grave warning of peril to the inhabitants of the Earth. The surges in the solar protons causing the Moon to brightly glow in red light will also expose populations to severely increased background radiation.

Chapter 4
Stars of Heaven Fall to Earth

> And the stars of heaven fell unto the earth, even as a fig tree casteth her untimely figs, when she is shaken of a mighty wind. (Revelation 6:13)

Section 4-1. Identifying the Falling Stars

Ellen G. White had argued this scripture had been fulfilled by the very dramatic Leonid meteor shower of November 12, 1833.[262] Eyewitnesses had compared its intensity with a snowstorm.[263] Rates of fall varied between a spectacular 50,000 to 200,000 meteors per hour.[264] People were terrified by it. Many thought the end of the world was imminent and began to study Bible prophecy. They did not flee to caves. For others, it was the beginning of modern meteor astronomy.[265]

The 1833 Leonid meteor shower did not fulfill all of the details given in Revelation 6:13 – it did not produce impacts. The *New Jerusalem Bible with Apocrypha* and *Darby's Bible*, stated the stars will fall onto the earth. *Young's Analytical Concordance to the Bible* indicates the word "unto," in the King James Version was translated from a Greek word that means "into." This prophecy explicitly predicted the Earth will be hit by falling points of lights. According to McKinley,[266] none of the meteors hit the ground! That statement is hard to believe. The 1833 meteor may not have produced craters and dramatic impacts, but the tiny meteors probably reached the surface.

The massive Leonid meteor shower of 1833 occurred over the eastern United States. It did not cause people to hide in caves and dens.

Survivors of the 1490 meteor shower may have hidden in caves and dens in China. It killed 10,000 people in Shanxi Province.[267] Neither one of those meteoric incidences caused the world population to flee to underground shelters.

The meteors of the 1833 shower did not fall like figs shaken from a wind blown tree. Some of the observers noticed the streaks of light – meteors – had emanated from the neck of the constellation of stars known as Leo. The painting of the meteor shower over Niagara Falls;[268] many woodcuts in magazines and newspapers;[269] and the description in Silliman's Journal[270] of 1833 indicated the streaks of light emanated from a single point in the sky.

Radial points, the principal feature of meteor showers, disqualify them as candidates for fulfilling this prophecy. A common point of emergence does not characterize the prediction of stars falling to the Earth in verse thirteen. The points of light are to fall like figs from a tree shaking in the wind. Figs fall from many points distributed throughout the branches of a tree. They can not fall from a single point of origin. The paths of the points of light described in Revelation 6:13 will not have a radial point. Therefore, meteor showers fail to satisfy all of the details given in Revelation 6:13.

A popular explanation that incorporates numerous falling incandescent objects that do have a radial point is nuclear war. It provides a chain reaction of convulsions stemming from the use of atomic bombs, thermonuclear warheads, and missiles in a future world war. One of the traits shared by such commentaries is the explicit references to weapon systems (e.g., Minuteman Intercontinental Ballistic Missiles (ICBM)) and the comprehensive descriptions of the effects of nuclear explosions.

Taylor stated the effects from the use of thousands of multi-megaton thermonuclear warheads would cause the Sixth Seal events.[271] He said the disruption of fault-lines by the nuclear detonations would cause the earthquakes and the dislocation of mountains. He attributed the darkening of the Sun and red color of the Moon to debris clouds, but could not treat it as literal, global blackness because of photosynthesis. He stated the plants had to survive to provide oxygen and total darkness would have been too devastating. Taylor attributed the appearance of stars falling from heaven to incandescent re-entry of warheads that had been launched by ICBM's, intermediate range ballistic missiles

(IRBM), or fast fractional orbit ballistic systems (FOBS). He attributed the heavens departing as a scroll to the temporary blast effects of the explosions. Taylor treated the flights to air raid shelters, caves, and rocks as an expected outcome of the war.

Hal Lindsey (1977) used fewer and less powerful nuclear explosions in his explanation.[272] He described the great earthquake and dislocation of mountains as a violent catastrophic shaking that will be caused by nuclear detonations. Debris from cobalt bombs was used as the cause of the blacken Sun and redden Moon. Lindsey also used the reentry of Russian FOBS warheads to explain the stars falling from heaven passage. And, the heaven departing as a scroll was attributed to the blast effects of nuclear explosions.[273]

The primary weakness of this type of explanation is the lack of surprise. Military facilities would be aware of missile launches hours before the darkening of the Sun. According to Revelation 6:12, the stars fall after the solar and lunar images have changed. The nuclear war model does not support all of the scriptural details.

The tone of Revelation 6:13 described points of lights, that were customarily in the night sky and had been abruptly dislodged from their normal states of motion to fall onto the Earth. A suitable explanation must satisfy these characteristics.

Seismically induced lights, known as earthlights, satisfy some of the requirements. The association of electromagnetic signals with earthquakes has gained increasing acceptance in the scientific community. For example, Asada, Baba, Kawazoe, and Sugiura (2001)[274] have identified a class of very low frequency signals that precede earthquakes with Richter scale Magnitudes between 4 and 6. More intense electrical phenomena have been reported prior to, during, and/or after earthquakes. Flavius Josephus[275] reported lightning leaping from fissures in the earth that were caused by loud earthquakes near Mizpeh in B.C 1120. Dr. Martin Altschuler cited descriptions of several Japanese earthquakes that had produced globes of light, blue fire, and ball lightning.[276] Peter Brookesmith cited several cases of earthlights that transpired in the vicinity of geophysical zones of exceptional seismic activity.[277] Reports of spherically shaped earthlight emissions have spanned several years in the Hessdalen Valley, Central Norway. Mainstream experts had dismissed Hessdalen phenomena reports until photographs taken from the ground and orbits firmly evinced their reality.[278] Since 1984, various instrumented observational investigations of the phenomena have been conducted in that area.[279]

Some earthlights have drifted to the ground during earthquakes. Revelation 6:13 states the lights will drop like figs from a tree. The slow drifting behavior of earthlights renders them unsuitable as candidates for fulfilling this scripture.

Examinations of the Greek words that were translated for the text yielded clues of the identity of the phenomena. The word star had a descriptive meaning during the time the book of Revelation was written. It was translated from *astor*, [SEC 792] which means an object that glimmers. The Greek word that was translated as heaven has been used to represent the sky – the domain of clouds. The question to be answered is what are the glimmering objects that are normally in the sky? One answer is lightning bugs. But they can not qualify as prophecy candidates. Falling lightning bugs will not terrify populations.

A thought experiment based on Revelation 6:13 could help us arrive at a solution. According to this scripture, the Sixth Seal is opened, a great earthquake occurs, the Sun is darkened, and the Moon becomes as blood. No matter where a person may be located, darkness covers the whole planet. On the night side of the planet, the Moon would become red. On the daylight side, the sky darkens. Points of lights that are normally found in the night sky are aircraft. The magnetic storms that will cause the Sun to blacken will also cause tremendous surges in the production of both electromagnetic radiation and solar energetic particles. Both could reach levels that would incapacitate aircraft propulsion systems and aircrews. The nightlights of those disabled flying machines would appear as stars falling to the ground.

Section 4-2. Wingless Aircraft

There is a second aspect of that scripture that needs to be examined. To understand it requires a little thought experiment. Imagine two, identical, unmanned airplanes flying in the same direction and altitude with the same speed. If both planes simultaneously lost power and only one of them lost all of its wings, which one would hit the ground first? The one without wings would fall more rapidly to hit the ground first. Wings slow the rate of descent. The fall of wingless aircraft would resemble the fall of figs. Figs are wingless. Revelation 6:13 indicates the stars will fall as figs shaken from a wind blown tree. We can assume those falling stars may be wingless aircraft dropping to the ground.

Let us consider the third feature of that prophecy. The current volume of world wide air travel seems to be too small for its abrupt loss to traumatize mankind. Acts of terrorism during September 11, 2001 have caused temporary reductions in commercial flights. Rising fuel prices have produced long term reductions in commercial air traffic. Fulfillment of this prophecy seems to occur after a substantial resurgence of that industry with wingless aircraft.

What kind of technology allows flight without wings? Helicopters and gyrocopters are a viable category. But, they have not enjoyed the mass production of fixed wing aircraft. Conventional trend analyses into advanced air propulsion systems predict the continued use of wings for aircraft.[280]

Experiments have been conducted with wingless aircraft that are usually disk shaped. They have been called aerodynes. Even though United States patents have been awarded for hundreds of aerodynes[281] from 1932 through 2001, none of them have significantly populated our skies.

Military attempts have been limited to experimental models. A declassified intelligence report described a proposal for developing turbojet powered aerodynes.[282] This project became known as Silverbug. It failed to attain its flight requirements, the program was cancelled, and the secret article was declassified. Hilton performed a study on the re-entry characteristics of saucer-shaped vehicles.[283] A proposal for a manned, orbit capable bomber, known as the lenticular reentry vehicle (LRV), was one of the products of the Pye Wacket Project. The General Dynamics Corporation[284] conducted a classified Pye Wacket Project feasibility study; Blanchard[285] performed an analysis of the landing characteristics of a LRV; and a classified report on the LRV was written by Oberto.[286] His report was declassified and cleared for public release on December 28, 2000 and was the foundation of the *Popular Mechanics* article by J. Wilson.[287] He used photographs of crash site debris to evince flight testing of a full-scaled LRV. If they had been operational, their mass production was probably too expensive to serve as the population of falling stars described in Revelation 6:13. Oberto's report indicated the LRV had used rocket propulsion systems.

Bulbous wedge shaped vehicles have been incorporated with re-entry tests. They resemble large gum drops with little, vertical fins (e.g., Martin Marietta X-24A and X24B and the Northrop/NASA HL-10 and M2F2). They do not fly. They were designed to expand the maneuvering capabilities of manned spacecraft that were re-entering from orbit. Level flying was not one of their attributes. Revelation 6:13 seems to depict commercial air traffic that is disrupted.

Research into developing vertical and/or short takeoff and landing (V/STOL) aircraft may have yielded a clue to the problem. Experimental aircraft with tilting, ducted fans had been created for V/STOL feasibility studies (e.g., Bell X-22A, Nord 500, and Piasecki VZ-8P (B) Airgeep II). They relied solely on ducted fans for lift. Subsequently, their high fuel consumption rates blocked further development for commercial use. Extant technology for such fans has not been able to compete against the cost effectiveness of winged vehicles.

Section 4-3. Wingless Aircraft Propulsion Systems

My search for propulsion systems that could satisfy all specific aspects of Revelation 6:13 was then shifted to finding published reports about manmade, self-contained, all weather functional aerodynes. "Levitating gifts" from angels, extraterrestrial beings, and/or time travelers were excluded. Also, excluded were theoretical propulsion systems. This restriction substantially reduced the number of exotic possibilities that had been cited in the web sites of Robert Stirnaman, Robert A. Nelson, Jean-Loius Naudine (JLN) Laboratory, the Gravity Society, and the formerly funded NASA Breakthrough Propulsion Physics Program.

The next restriction was to make the selection from literature about self contained, working models. For example, the spectacular, diverse, and numerous lifters inspired by JLN Laboratory had to be omitted because they currently lack the strength to lift their own power supplies – they were not self-contained. Subsequently, speculations applied to the devices that remain shall have a higher probability of satisfying the Revelation 6:13.

The second set of restrictions was applied to the published

descriptions about candidate propulsion systems. Peer reviewed papers were treated as the optimal documents. Reported performance had to follow replications of requisite propulsion characteristics. The replication prerequisites had to be within the capabilities of extant technology. And, the current theoretical foundations had to support and/or predict the performance characteristics of the propulsion candidate.

The two sets of selection criteria were used for identifying potential candidates for fulfilling the falling stars prophecy. Gary Bennett's set of more stringent criteria[288] would have to be satisfied by the actual device before its mass production, marketing, and global deployment.

The movies may have provided a glimpse of the technology that may revolutionize air travel. Cell phones, directed energy weapons, satellite television dishes, cloned animals, portable radios, and rocket powered spacecraft had appeared first in science fiction. Recently, a new family of jet engines had begun to appear, with little fanfare, in science fiction films. The fictional engines have enabled platforms and weapon systems to casually hover for hours, like helicopters. In the real world, commercial and military jet engines consume too much fuel to permit lengthy loitering times. The movie jet engines were virtually propellantless. The overt emergence of such systems could revolutionize the aviation industry. The reduced fuel requirements would rejuvenate air travel.

Reports of such engines appeared near the close of the Second World War. Documents and eye witness accounts about its performance claimed it was able to compress air through an implosion process that produced explosive releases of energy without the use of propellants. The energy generated by the change in states provided the suction power to continuously draw in air and release it in a manner like a turbine. Since its fuel was only air, it needed only a starter motor to accelerate the bladeless turbine waveform plate to 10,000 rpm.

You may wonder why it had not been used by the aircraft industry. The answer is the vortex turbines that Viktor Schauberger had constructed for flight had been confiscated at the close of World War II. Unlike the other German secret weapons, its surviving documents and models remained concealed for decades.

A couple of researchers have recently verified the reality of Viktor Schauberger. After Viktor had died in 1958 his son, Walter, began to release material about his father's work to Callum Coats.[289] The releases were performed by Walter a few years before his death in 1997. Coats translated and compiled Viktor's writings and patents and published them in four volumes. The first volume published[290] in 1996 featured Schauberger's philosophy and general achievements. The fourth[291] volume was a compilation of translated technical articles. Nick Cook,[292] former aviation editor of *Jane's Defense Weekly*, visited the Schauberger Institute in Bad Ischl, Austria, interviewed Viktor's grandson, Joerg Schauberger, and examined their stacks of documents.

Subsequently, several web sites have featured photographs and descriptions of Schauberger's achievements. Frank Germano's web site, "Viktor Schauberger; the Water Wizard," featured a biography, numerous quotes from Viktor's writings, diagrams from his patents, and technical reports by Guy Letourneau's tests that had been conducted by International Turbine & Power, LLC on similar turbines – the Tesla pump. Jean-Louis Naudin's web site, JLN Labs, incorporated excellent photographs and diagrams illustrating the flow and compression of air. They provided performance descriptions of Schauberger's January 1940 and 1941 turbines; the Repulsins Type A and B, respectively. Naudin's web site carried five reports by James L. B. Bailey on the hazards of constructing and testing Repulsins. Bailey gave several strong warnings about the ability of Repulsin turbines to produce very high voltages and hard x-rays!

Other authors have cited Callum Coats' book that carried photographs of one of the Repulsins.[293] Stevens' book[294] contained the testimonies of individuals who had worked on components of the Repulsins for the Third Reich. It featured an awesome estimate made by Schauberger of the lift-to-weight ratio that had been produced by a Repulsin. It had torn off its anchor bolts to fly into the ceiling. Coats' translation of the fourth volume of the echo-technology series featured photographs of a Repulsin that Karl Gerchsheimer had given to Commander Richard C. Feierabend, U. S. Navy (retired), during October, 1944.[295]

If the Repulsins can be replicated and perform as well as the descriptions from the 1940's reports, air travel could have the resurgence that has been deduced from Revelation 6:13. The heart of the Repulsins has not been exploited by current technology, but its components yielded clues to its operating principles.

Adhesion and viscosity seem to have been one of the properties used by Schauberger's bladeless turbine to draw in air. They have been effective in simpler variations of his turbines. Nikola Tesla had exploited these principles with closely spaced, parallel plates in his pumps and bladeless turbines.[296] Germano, Dorantes, Johnson, and Letourneau had manufactured stainless steel Tesla turbines that handled steam, water, propane, and compressed air.[297] A series of promising tests have been conducted on the Tesla pump as an artificial heart by university biomedical engineering programs in Quebec, Texas, and Virginia.[298] Adhesion and viscosity could be the primary principals of the first stage of compression for Schauberger's turbine.

Technical, peer reviewed journals have yet to publish articles on the effects claimed by Schauberger. The language and explanations in his publications did not incorporate the conventions of engineering and physics. If the recently released historical documents and artifacts are valid, Schauberger's turbines could be an excellent candidate for revolutionizing future air travel. Explaining Revelation 6:13 in terms of falling, wingless aircraft, accommodates current and future propulsion technologies.

Section 4-4. What Causes the Points of Light to Fall

The question becomes, what kill mechanism knocks the aircrafts from the sky? How could solar storms immediately incapacitate aircraft power plants, controls, and/or aircrews? According to popular media and science fiction, the candidate for the aircraft kill mechanism should be the electromagnetic pulse (EMP). Various web sites, articles, and media have claimed high altitude detonations of atomic weapons would cause wide area damage to electrical systems.[299] But, careful examinations of nuclear test reports and rigorous analytical modeling have manifested the over exaggerated destructive capabilities of a category of EMP called TEMP's.

EMP effects, also known as "radioflash," were first observed on July 16, 1948 during the Trinity atomic bomb test in Alamogordo, New Mexico. Allegedly large range EMP effects were first reported from the

1962 Starfish high altitude hydrogen bomb tests. EMP's from the low and high altitude nuclear explosions were produced by the same mechanism. Collisions between the very intense flash of gamma rays, produced by the nuclear detonation, with the electrons orbiting the air molecules instantly causes a rapid EMP. Such an EMP has been referred to as the tachy EMP (TEMP). A magnetohydrodynamic (MHD) EMP occurs a brief moment after the TEMP and runs for almost two minutes.

Aircraft crashes have not been caused by electrical damages stemming from EMP's produced by nuclear weapons tests. EMP's from the atomic explosions over Hiroshima and Nagasaki did not cause the bombers to crash. According to Rabinowitz, the electrical failures at Oahu did not stem from the 1962 Starfish test.[300] That test involved the detonation of a hydrogen bomb 248 miles above Johnston Island and over 800 miles away from the Hawaiian islands.

The conversion of energy from a nuclear explosion into a TEMP is insufficient for long range damaging effects. Only one millionth of a bomb's energy is converted into a TEMP.[301] Analyses by Vittitoe and Rabinowitz manifested large retarding forces as the responsive mechanism that effectively mitigates effects caused by TEMP's[302]. Long range TEMP effects produced by lightning are more powerful than those from a nuclear explosion. The effects of solar storm, geomagnetically induced currents (GIC) are similar to the MHD EMP of nuclear weapons.[303] MHD EMP's are the second category of EMP.

According to Plait, mainstream astronomers have accepted the damaging capabilities of an EMP produced by gamma rays from distant celestial sources.[304] LaViolette's dissertation research led him to predict the existence of galactic core explosion phases that could produce burst of gamma rays strong enough to disrupt electrical systems on Earth.[305] Additional research yielded evidence for several spiral galaxies experiencing core explosion phases.[306] Mainstream astronomers did not embrace LaViolette's deductions due to the belief that gamma rays came from sources located within the Milky Way Galaxy.

Eleven years later, those beliefs were destroyed. On August 27, 1998, a record setting wave of gamma rays swept through our solar system. It triggered gamma ray detectors on board the flotilla of space probes scattered throughout the space between the Sun's planets and radio blackouts for various areas on Earth.[307] The wave hit the night side of the Earth at 3:22 Pacific Day Light Time and caused the ionosphere to drop approximately twelve miles for five minutes.[308] Astronomers used the arrival times of the gamma ray wave collisions with the space probes to triangulate the location of its source. Its identity was a neutron star by the name of SGR 1900+14. Its location was 20,000 light years away from the Sun – placing it beyond the Milky Way Galaxy. This event evinced the existence of extra-galactic sources of gamma rays. Neither it nor subsequent outbursts produced EMP phenomena.

Analyses of such outbursts of energy may provide additional insight into the significance of the sackcloth appearance of the blackened Sun. Coburn and Boggs[309] claimed to have detected 80% linear polarization within the gamma rays emitted by the gamma ray burst of December 6, 2003 from GRB021206. The high polarization values indicated the rays had been produced by very structured and powerful magnetic fields. It had been assumed that the explosions producing the gamma ray bursts had been accompanied with chaotic and disorganized magnetic fields.[310] Rutledge and Fox have challenged those measurements.[311] If the controversy is resolved in support of Coburn and Boggs, several gamma ray burst theories would have to be discarded.

Boggs, Coburn, and Kalemci were the first to perform polarimetry on the gamma ray emissions from two solar flares.[312] Their analyses of the X-class solar flares of July 23, 2002 and October 28, 2003 were approximately 21% and -11% polarizations, respectively. Although the polarization levels were moderate, they implied their source was structured phenomena within the flares. Analyses by Hurford, et. al. of the gamma rays from the X-class solar flares of October 28, 29, and November 2, 2003, indicated the gamma rays had been emitted from compact sources within the flare active regions.[313]

Polarization attributes of gamma rays implied their creation was a product of structured magnetic fields. The sackcloth image of the blacken Sun denotes a structured environment. It is reasonable to expect such processes to generate very high fluxes of very energetic particles and gamma rays. The gamma rays may not produce fatal TEMP's, but other mechanisms that could incapacitate aircraft were examined in Chapter 5.

The Seven Signs of the Apocalypse, a religious commentary telecasted by the History channel, used the possibility of future gamma ray bursts from the Wolf-Rayet star, WR104, to explain Revelation 6:12. Astronomers expect WR104 to explode into a supernova within the next few hundred thousand years. Distance estimates for WR104 are from 5,000 to 8,500 light years away from the Sun. If certain conditions occur when it explodes into a supernova, intense beams of gamma rays could blaze forth along its axis of rotation. WR104's axis of rotation is pointing in the general direction of the Earth. *The Seven Signs of the Apocalypse* commentators indicated the gamma rays from that event could cause the formation of thick, reddish brown smog in the upper atmosphere. They said it would cause the darkening of the Sun in the manner predicted by Revelation 6:12. We could assume that mechanism could cause aircraft to drop from the sky by incapacitating the aircrews.

Obscuration is the major shortcoming of the WR104 explanation. Dark smog would not yield a textile image of the Sun black as sackcloth of hair. The global winter caused by the gamma ray induced smog would not cause populations to seek shelter beneath cliffs. Cliffs offer very little protection from severe drops in temperature. Hiding in caves and dens would be the primary outcome for survivors. Intensified gamma radiation does not yield fulfillment of every aspect of the events stemming from the opening of the Sixth Seal.

Chapter 5
Heaven Rolls Back

And the heaven departed as a scroll when it is rolled together; and every mountain and island were moved out of their places. (Revelation 6:14a)

The word heaven in this passage denotes the sky. This segment of the verse is saying the sky rolls away like a scroll. I do not believe this is a description of the removal of Earth's atmosphere. Such an event would expose regions of the Earth to the vacuum of space. That would cause massive fatalities. The sixth chapter of Revelation does not refer to deaths associated with the opening of the Sixth Seal.

The Earth's atmosphere does expand and contract with solar variability. According to Yenne, that fact was brutally encountered by America's first space station, Skylab.[314] The 100-ton vessel had been placed into a 270-mile orbit on May 14, 1973. Increased sunspot activity from 1978 to 1979 caused Earth's atmosphere to expand and, through atmospheric friction, Skylab's speed gradually decreased. Satellites must maintain their speeds to remain in orbit. The occasional expansion of the Earth's atmosphere reached high enough to slow down Skylab. It fell out of orbit like a fiery meteor and hit the Indian Ocean on July 11, 1979. The atmosphere may expand during this scripture segment, but this type of disturbance would not look like the sky parting.

Analyses by Palle, Butler, and O'Brien yielded a positive correlation between cosmic rays and low level cloud cover.[315] Lag times between solar activity and climatic response have been identified by Perry for various regions of the world and vary from 25 to 70 years.[316] During the six seal solar blackout, low altitude clouds may thicken or diminish with corresponding increases or decreases in particle energies from the Sun. This process could be an explanation for the rolling pattern described in this scripture segment. But, the Sun would be too dark to make the phenomenon visible from the sparsely populated areas. Another phenomenon that occurs in the night sky may provide the dramatic images resembling heaven rolling back like a scroll.

What would the sky look like as it rolls back like a scroll? We could compare it with a carpet rolling away from a small, flea sized observer, watching it from the floor. The flea would see the threads of the carpet reach high above it to curve back and recede away. A broad row of tall lines in the sky could yield a similar view. In the night sky, those lines would have to glow. Can a solar storm produce auroral displays that shimmer across the entire night sky of the globe? The answer is yes. During the era of the telegraph, a few years before the beginning of America's Civil War, the Earth was bathed in dazzling aurorae. Newspapers, magnetic instruments, and telescopes around the world recorded the great auroral phenomena. It was not the first record of auroral activity. It was the first auroral storm to generate documents that connected sunspots, solar flares, auroral activity, and electrical Earth currents. This was a major milestone for a new era in the studies of Sun-Earth connections.[317]

Section 5-1. The Great Auroral Storm of 1859

Two worldwide auroral displays took place within four days of each other. The first one occurred on Sunday, August 28th and second one on Thursday night and Friday morning of September 1st -2nd of 1859. The second auroral storm was brighter, longer, and more extensive than the first. Witnesses were awed by the dazzling streamers and curtains of light dancing across the sky. It was a fantastic demonstration of the Sun-Earth connection.

James Green, et al, published a sample of accounts from numerous newspaper articles and reports about both displays from cities and ships scattered about different latitudes.[318] Additional auroral accounts and histories of solar science were in books by Whitehouse[319] and Clark[320] and in articles by Cliver[321] and Clark.[322] All four histories included dramatic descriptions of the problems telegraph stations had with dangerous surges in electrical current and disruptions to transmitting messages. There were a few incidents where equipment had burst into flames and telegraphers were shocked. One telegrapher had been shocked into unconsciousness during the September auroral storm. The accounts compiled by Cliver and Clark provided glimpses of how the public, around the world, interpreted and responded to the spectacular phenomena.

Newspaper editors and reporters attributed the problems of the telegraph stations to electrical properties of the auroras. The Great Auroral Storm of 1859 displayed an electrical connection to auroras that could not be explained in terms of reflections off of a lake. Prior to the Auroral Storm, upper atmospheric lightning, light reflected from icebergs, nebulous matter, meteors, and debris from volcanoes were used to explain auroras.[323] It would take astronomer over seventy years to identify and prove the connection between auroral phenomena and earth currents.

None of the four auroral histories, cited above, featured descriptions of the sky rolling away like scrolls. But, the visual elements necessary to convey such an image were seen around the globe and in both hemispheres. There were descriptions of colored streamers jetting high into the sky; shimmering curtains of lights; and lights arching overhead. Such features may vividly appear in the future sky caused by the darkened Sun to convey the appearance of heaven rolling away like a scroll. Analyses of the newspapers and reports by the American mathematician, Elias Loomis, indicated the auroral displays were components of a large ring that started fifty miles above the ground and had stretched up to 500 miles in space.

Section 5-2. The Carrington Event

The cause of the Great Auroral Storm had occurred almost eighteen hours earlier and had been independently observed and recorded in England.

On September 1, 1859, at 11:15 AM something like the faint sound of a footstep by a distant tyrannosaurus was recorded in dark rooms in the Kew Observatory and the Greenwich Observatory, England. No one heard or felt it. It was recorded by beams of light shifting away from their normal positions that shined on photographic papers attached to slowly rotating drums. The arrays of instruments were self-recording photographic magnetographs. They measured the variations in the vertical, horizontal, and declination components of the Earth's magnetic field. Traces of the declination and horizontal components were recorded on the same photographic sheet.[324] The magnetograph for the horizontal component of the field recorded a five-minute deflection with maximum amplitude of approximately 110 nano-Teslas (nT). Back then, the hump was called a magnetic crochet because of its hook-like shape on graphs. Crochets appeared in both the declination and horizontal components. But, the synoptic analyses by Cliver and Svalgaard indicated the magnetic crochet in the horizontal component was among the largest ever reported.[325] Other magnetic stations had failed to record it. They had not detected it because their self-recording photographic magnetographs had been off or their practice of taking readings from their magnetometers every hour on the hour had caused them to miss it. Crochets were distilled by retrospective analyses of data from the observatories in St. Petersburg, Barnaul, and Nerchinsk, Russia.[326]

The reliability of Kew's magnetic instruments had been certified May 8, 1860 by one of the fathers of the "Magnetic Crusade," Edward Sabine.[327] Problems encountered during their first year of operation, in 1858, had been effectively corrected by the installation of a water regulator to govern the flow of gas to the lamps. The amazing magnetic crochet had been accurately recorded.

The Great Auroral Storm occurred 17.6 hours after that magnetic crochet. Magnetic crochets are now known as solar flare effects (SFE). SFEs are a form of sudden ionospheric disturbance caused by torrents of X-rays and ultraviolet light that have erupted from a solar flare to strike the ionosphere. Electrical ring currents are generated in the upper atmosphere by those collisions and cause changes in the Earth's magnetic field. The SFE of 1 September 1859 was tremendous. At the time, its significance was not realized because it had been dwarfed by extreme readings recorded during the Great Auroral Storm.

The flares that had caused the SFE of 1 September 1859 had been observed by two Fellows of the Royal Astronomical Society, Richard Christopher Carrington and Richard Hodgson. From July through September 1859, a large sunspot group had been under observation.[328] At 11:05 AM, local time, on 1 September 1859, they saw dazzling white lights suddenly appear among the sunspots. Both of them noted their observations were simultaneous with the time the magnetic instruments at Kew Observatory had been disturbed.

Richard Hodgson observed the entire event from his home in Highgate without interruption. The following was his November 11, 1859 report to the Royal Astronomical Society:

> While observing a group of solar spots on the 1st September, I was suddenly surprised at the appearance of a very brilliant star of light, much brighter than the sun's surface, most dazzling to the protected eye, illuminating the upper edges of the adjacent spots and streaks, not unlike in effect the edging of the clouds at sunset; the rays extended in all directions; and the centre might be compared to the dazzling brilliancy of the bright star α *Lyrae* when seen in a large telescope with low power. It lasted for some five minutes, and disappeared instantaneously about 11.25 A.M. Telescope used, an equatorial refractor 6 inches aperture, carried by clockwork; power, a single convex lens, 100, with a pale neutral-tint sunglass; the whole aperture was used with a diagonal reflector.
>
> The phenomenon was too short duration to admit of a micrometrical drawing, but an eye-sketch was taken, from which the enlarged diagram has been made; and from a photograph taken at Kew the previous day, the size of the group appears to have been about 2^m 8^s, or (say) 60,000 miles.
>
> The magnetic instruments at Kew were instantaneously disturbed to a great extent.[329]

Richard Carrington was able to observe more details about the event than Hodgson because his projected images onto a screen of regions of the Sun were eleven inches in diameter. And, they had a superimposed crosshair over the solar image to facilitate precise drawings of sunspots. Carrington recorded timings from a very accurate chronometer he had installed in the observatory attached to his manor in Redhill. The following was extracted from his November 11, 1859 report and their diagrams to the Royal Astronomical Society:

> While engaged in the forenoon of Thursday, Sept. 1, in taking my customary observation of the forms and positions of the solar spots, and appearance was witnessed which I believe to be exceedingly rare. The image of the sun's disk was, as usual with me, projected on to a plate of glass coated with distemper of a pale straw colour, and at a distance and under a power which presented a picture about 11 inches diameter. I had secured diagrams of all the groups and detached spots, and was engaged at the time in counting from a chronometer and recording the contacts of the spots with the cross-wires used in the observation, when within the area of the great north group (the size of which had previously excited general remark), two patches of intensely bright and white light broke out, in the positions indicated in the appended diagram by the letters A and B, and of the forms of the spaces left white. My first impression was that by some chance a ray of light had penetrated a hole in the screen attached to the object-glass, by which the general image is thrown into shade, for the brilliancy was fully equal to that of direct sun-light; but, by at once interrupting the current observation, and causing the image to move by turning the R.A. handle, I saw I was an unprepared witness of a very different affair. I thereupon noted down the time by the chronometer, and seeing the outburst to be very rapidly on the increase, and being somewhat flurried by the surprise, I hastily ran to call some one to witness the exhibition with me, and on returning within 60 seconds, was mortified to find that it was already much changed and enfeebled. Very shortly afterwards the last trace was gone, and although I maintained a strict watch for nearly an hour, no recurrence took place. The last traces were at C and D, the patches having travelled (sic)

All Shall Hide

considerably from their first position and vanishing as two rapidly fading dots of white light. The instant of the first outburst was not 15 seconds different from $11^h\ 18^m$ Greenwich mean time, and $11^h\ 23^m$ was taken for the time of disappearance. In this lapse of 5 minutes, the two patches of light traversed a space of about 35,000 miles, as may be seen by the diagram, which is given exactly on a scale of 12 inches to the sun's diameter. On this scale the section of the earth will be very nearly equal in area to that of the detached spot situated most to the north in the diagram, and the section of *Jupiter* would about cover the area of the larger group, without including the outlying portions. It was impossible, on first witnessing an appearance so similar to a sudden conflagration, not to expect a considerable result in the way of alteration of the details of the group in which it occurred; and I was certainly surprised, on referring to the sketch which I had carefully and satisfactorily (and I may add fortunately) finished before the occurrence, at finding myself unable to recognize any change whatever as having taken place. The impression left upon me is, that the phenomenon took place at an elevation considerably above the general surface of the sun, and, accordingly, altogether above and over the great group in which it was seen projected. Both in figure and position the patches of light seemed entirely independent of the configuration of the great spot, and of its parts, whether nucleus or umbra.

It has been very gratifying to me to learn that our friend Mr. Hodgson chanced to be observing the sun at his house at Highgate on the same day, and to hear that he was a witness of what he also considered a very remarkable phenomenon. I have carefully avoided exchanging any information with the gentleman, that any value which the accounts may possess may be increased by their entire independence.

(Mr. Carrington exhibited at the November Meeting

of the Society a complete diagram of the disk of the sun at the time, and copies of the photographic records of the variations of the three magnetic elements, as obtained at Kew, and pointed out that a moderate but very marked disturbance took place at about $11^h\ 20^m$ A.M., Sept. 1st, of short duration; and that towards four hours after midnight there commenced a great magnetic storm, which subsequent accounts established to have been as considerable in the southern as in the northern hemisphere. While the contemporary occurrence may deserve noting, he would not have it supposed that he even leans towards hastily connecting them. "One swallow does not make a summer.")[330]

The Carrington Flare

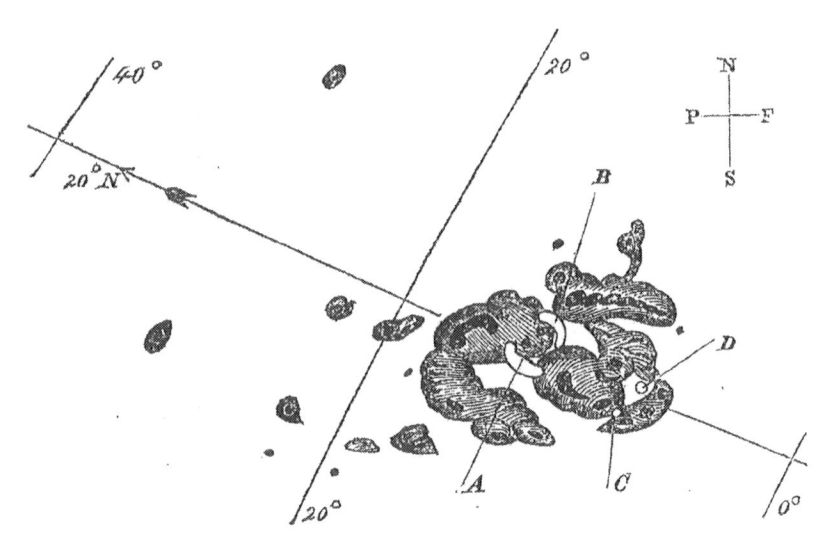

Figure 2. The scaled diagram drawn by Richard Carrington of the sunspot group and white light solar flares he observed September 1, 1859. They suddenly appeared at regions A and B and disappeared at C and D, respectively. The two patches of light traveled 35,000 miles, in 5 minutes (420,000 miles per hour).

A few months later, during the total solar eclipse of July 18, 1860, astronomers saw the mechanism that conveys significant portions of the magnetic energies from solar active regions to the Earth: a coronal mass ejection (CME).[331] Reports from observers along the path of totality reported the formation and ascent of a glowing bubble from the Sun.[332] Two photographs of the event were interpreted as a prominence originating from the Sun, not the Moon.[333] At that time, astronomers did not realize the significance of the events they had painstakingly recorded. We now know that solar active regions are populated with sunspots. Those regions provoke the emission of tons of plasma known as CMEs. The particles in the CME cause auroral displays. Magnetic lines of force of the CMEs perturb the Earth's magnetic field as first evinced by the erratic behavior of compasses increasing and decreasing as the number of sunspots grew and declined. Therefore, it is reasonable to expect the phenomena associated with blacken Sun of Revelation 6:12 to be considerably more intensified than the effects of the Carrington Event

Correlations between CMEs and earthquakes suggested the expectation of records of seismic activity transpiring within a couple of years of the sunspot group that hosted the Carrington Event. A search of the literature did yield positive results. Earthquakes occurred in 1859 on August 10th in San Francisco and San Jose, California; August 21st in Padstow, England; September 22nd and October 5th in San Franscisco;[334] October 22nd in Norcia, Italy;[335] and December 1st and 24th in San Francisco. In 1860, they occurred in Charlevoix, Canada; January 13th in Newquay, England; April 8th in Haiti,[336] along the coast from Petit-Groave to Anse-à-Veau;[337] April 16th and September 23rd in San Francisco and Martinez; October 23rd in the Caribbean.[338] Future analyses of these may distill a causal link with the geomagnetic storms of the Carrington Event. This would be consistent with the effects associated with the crucifixion of Jesus Christ and the prophecies described in Revelation 6:12, 14.

Section 5-3. Ice Cores

Several decades elapsed before astronomers understood the connection between solar active regions and the behavior of compasses, electric currents in the ground and conductors (e.g., oil pipes and power lines), and auroral displays. Many of those geomagnetic storms were produced by tons of plasma from the corona known as coronal mass ejections (CMEs). A typical CME is composed of twenty billion tons of plasma from the corona that travels at several miles per second.[339] Solar flares would cause geomagnetic storms less frequently than CMEs. Both are sources of very powerful energetic particle events.

Such events had occurred in 1864, 1878, 1894, 1895, 1896, 1960, 1972, and 2000. Some of them had been given special names. For example, the energetic particle event of July 14, 2000 was named the Bastille Day event. Some of these had struck awe in the scientific community.

The power of the1972 event had been alarming for the planners of the Apollo 17 mission to the Moon. Apollo 17 was successfully launched in December 7th; the last manned activity on the Moon transpired December 13th; and splashdown of the Command Module in the Pacific Ocean and the recovery of its astronauts were completed on the 19th. But, on August 4, 1972, an anomalously large solar proton event produced an unshielded radiation dose of 20,000 REM above the Earth.[340] The average lethal dose has been defined to be 450 REM. The ramifications of the 1972 solar proton event had been terrifying for the Apollo Program mission planners. This event and the damaging effects others had on communications satellites, transformers, and power grids stimulated research into the causes and effects of solar storms. To anticipate the upper limits of those storms, researchers developed techniques to measure the intensity of solar energetic particle events that had occurred before the Space Age.

It is now known that when the energetic particles from those events collided with Earth's upper atmosphere, complex chemical processes have caused the formation of various nitrates (NO_y). The nitrates have fallen slowly through the air to be absorbed by processes at the surface. Nitrates falling into Polar Regions have become a part of the ice pack. As layer after layer of snow fell, concentrations of nitrates would vary over the length of an ice core that had been drilled out.

A team of scientists from Kansas pioneered the science of extracting records of intense solar activity from ice cores. Edward J. Zeller, Gisela A. M. Dreschhoff, and Claude M. Laird from the Space Technology

Center, University of Kansas, showed in 1986, that impulsive nitrate concentrations in a glacial pit and a firn (very compacted snow) ice core from the Ross Ice Shelf, Antarctica, corresponded in time with the 1972 and 1984 major solar events.[341] In 1990, Dreschoff and Zeller published the correlated timings of the highest impulsive nitrate concentrations that had been contained in ice cores from the Windless Bight drill site on the Ross Ice Shelf, Antarctica, with the solar flare of 1928 and the solar proton events of 1946 and 1972.[342] In 1993 and 1999, Margaret Shea, D. F. Smart, Dreschhoff, and Zeller presented the correlations between large nitrate concentrations in polar ice cores and major solar proton events and/or solar terrestrial events.[343] In 2001, McCraken, Dreschhoff, Zeller, Smart, and Shea showed the probability of attributing the one-to-one correspondence between impulsive nitrate concentrations and the largest solar proton fluence events was one out of 870 million.[344] And, Table 1 of their paper indicated the year of the highest nitrate concentration and corresponding largest solar proton fluence (18.8 x 10^9 cm^{-2} protons with >30 MeV) was 1859 – the Carrington Event! That event had produced a fluence more than three times larger than that of the August 1972 event (5 x 10^9 cm^{-2} protons with >30 MeV). It had the highest fluence of the seventy largest impulsive nitrate events that had transpired between 1561 and 1950.

The ASDE stemming from Christ's opening of the Sixth Seal will produce solar particle fluences and energies that will exceed the Carrington Event. This hypothesis is based on the assumption that the weave textured blackening of the solar disk shall be much stronger than the sunspot group that had spawned the Carrington Event. To quantify the astrophysical (e.g., lunar luminescence) and geophysical implications of Revelation 6:12-17, one could examine the outcomes of intensifying the parameters that have been deduced from studies of the Carrington Event.

Further ice core research may strengthen evidence for ASDEs. During the 2002 PACLIM Conference, Dreschhoff presented graphs of impulsive nitrate concentrations that had been extracted from the ice cores of the South Pole, representing a time scale of 3,200 years, and ice cores from the south geomagnetic pole at Vostok Station, representing 1,200 years.[345] The two graphs in Figure 3 of her conference paper contained spikes that approximated the years of the thirteen ASDEs that had been presented in Section 2-5 of this book.

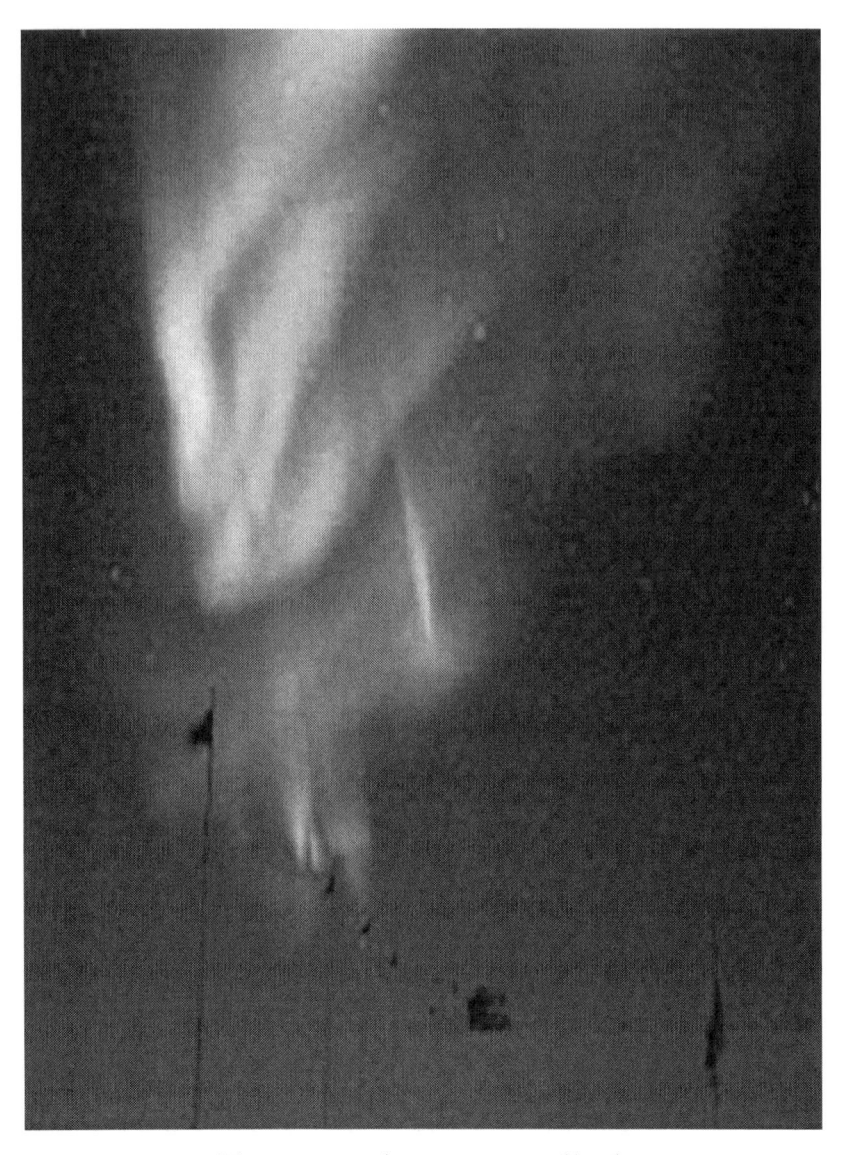

Figure 3. The space weather storm caused by the opening of the Sixth Seal event will cause auroral displays at the mid-latitudes much more dramatic than the one pictured above from the National Science Foundation Amundsen-Scott South Pole Station. It will look like a huge scroll rolling up.

Figure 4. Above is the view from space of an aurora.

Chapter 6
Mountains Are Moved

And the heaven departed as a scroll when it is rolled together; **and every mountain and island were moved out of their places.** (Revelation 6:14b)

Section 6-1. Physical Implications of Shifted Mountains

A review of the literature about this scripture yielded writers who used meteor impacts, nuclear explosions, and/or volcanic eruptions to explain the shift in locations of mountain and islands. Such mechanisms would cause high death tolls and global debris clouds that would expose mankind to the risks of extinction. This chapter presents a model that allows dislocations without the high loss of life.

The principle characteristics of the Sixth Seal earthquake are the dislocations of every mountain and island. A way to interpret this passage is to expect the distances between cities, not located in mountains or islands, to remain the same and to assume the distances between the peaks of mountains and islands to be changed. This scripture does not indicate the intensity of dislocation. It does not indicate if they will be dislocated vertically or horizontally by a fraction of an inch or miles and/or rotated and/or tilted by several degrees. No matter how they are "moved out of their places," we know the foundations beneath the mountains and islands will have to be modified to make it happen. The dislodgements may transpire briefly during the twelfth verse and receive geologic confirmations days later.

The intervening events may represent the time it takes geologists to discover and verify the changed locations. This passage evinces the survival of various technologies to allow measurements of all mountains and islands prior to the global migrations of all populations to dens and the rocks of mountains. The Global Position System has enabled surveyors to precisely measure vast geological features quickly.

The word 'every' was translated from the Greek word *pas* [SEC 3956]. It has been used to denote all, as in each and every type of category. Just having one dislodgement of a mountain or island, large or small, would be spectacular. If the above passage means every type of mountain and island is dislocated, at the very least the dislocations could be limited to one region of the globe. But, to incite worldwide migrations to underground facilities, the dislocations are likely to be global.

Approximately 24% of the surface of the Earth is mountainous. Islands are the tips of mountains based beneath oceans and seas. It is difficult to even begin to consider what it would be like to have all mountains and islands moved from their places. The Rocky Mountains, Mount Everest, Himalayas, Mount Kilimanjaro, Mount Ararat, Canary Islands, Hawaii, Singapore, Jamaica, and Puerto Rico are just a few of the world's mountains and islands that will be moved out of their places. Astronomical instruments atop mountains and maritime commerce of major islands could be adversely affected. To dislodge mountains differently from the cities shaken by the Sixth Seal earthquake implies there is a difference between the foundations that support them.

Some mountains are volcanoes. Beneath them are large lava chambers. The lava is a better conductor of electricity than the surrounding crust. "Volcanic cycles are said to coincide with sunspot cycles, although there is wide disagreement on this subject."[346] Intense electrical currents induced by the magnetic storms of the Sun may have caused some volcanic activity. Sunspots are associated with magnetic storms. If mountains that were not classified as volcanoes rested on substances that were excellent conductors of electricity, the dislodgements predicted in Revelation 6:14 could be caused by electrical currents within the Earth that were induced by the magnetic storms associated with the darkening of the Sun.

Section 6-2. Heliophysically Triggered Dislodgements

The source of power for moving the mountains and islands may be huge electromagnetic impulses that are emitted during the solar blackout. Recent studies have shown that some strong variations in the solar magnetic field, conveyed by massive coronal mass ejections (CME), have subjected our planet to stresses that have produced earthquakes. Several phenomena have been the product of reactions of the Earth to variations in the interplanetary magnetic field.

The motion of conductors through magnetic fields is resisted by the induced eddy-currents. Metallic artificial satellites have experienced spin-breaking because of the Earth's magnetic field. For example, one of America's earliest satellites, Vanguard I, slowed to a thirtieth of its initial spin rate within a two-year period.[347] The spin-down could have been prevented if Vanguard I had produced its own magnetic field to counter the influence of the geomagnetic field.

The Earth has its own series of electrical currents that help it maintain its spin rate. Seismic research has concluded the Earth's inner core is a mass of solid iron that is rotating slightly faster than the rest of the planet.[348] The molten outer core interacts with the inner core to generate currents of electricity and magnetic forces. These help the Earth resist a diverse variety of spin-down forces. Tidal effects caused by the gravitational fields of the Moon and Sun and the global friction between the surface and masses of water and air are the dominant forces resisting the Earth's rotation. According to the U. S. Naval Observatory, tidal friction causes a slowing of the length of days between 0.0015 and 0.0020 seconds per day per century. Stephenson had documented the long term changes in the length of days through examinations of historic writings about the dates and times of solar and lunar eclipses from Europe, China, Babylon, East Asia, and Arab countries.[349] Kokus listed several studies on the relationships between the Earth's rotation and solar activity.[350]

Section 6-3. Biblical Indications of Water Beneath Mountains

How can the mountains be moved? Is there a biblical precedence for this concept? The following scripture could be a hint for the answers for these questions:

> And every island fled away, and the mountains were not found. And there fell upon men a great hail out of heaven, [every stone] about the weight of a talent: and men blasphemed God because of the plague of the hail; for the plague thereof was exceeding great. (Revelation 16:20-21)

The twentieth verse indicates the islands shall flee. A variety of biblical translations treat this as islands rapidly moving away in a manner that suggests flights to safety. Since islands are mountains submerged in water, we can assume the phrase, "...the mountains were not found" implies the mountains had quickly disappeared from view, like the islands, or had been flattened. In either case, Revelation 16:20 indicates their dislocations will be very dramatic.

The twenty-first verse states an exceedingly great hailstorm occurs with or soon after those massive dislocations. Talents were used to express the weight of the hailstones. A talent is equivalent to approximately a hundred pounds! Can you imagine the effects of hundred-pound hailstones hitting the ground? Cars, buildings, storage tanks of grain and oil, refineries, and bridges would be pulverized into rubble that would be rapidly buried in ice.

Revelation 16:20-21 suggests a connection between the massive dislocations of mountains with turbulent water in the form of unusually violent hailstorms. The reoccurrence aphorism, Ecclesiastes 1:9, implies great hailstorms accompanying mountain and island displacements should have occurred earlier. According to the Hydroplate Theory by Walt Brown, such a phenomenon occurred thousands of years ago. We know of it as Noah's flood. He argued the critical role of water was a lubricant for the mountains.

Dr. Brown had developed his Hydroplate Theory to explain Noah's flood in geophysical terms based on the following scripture:

> In the six hundredth year of Noah's life, in the second month, the seventeenth day of the month, the same day were all the fountains of the great deep broken up, and

the windows of heaven were opened. And the rain was upon the earth forty days and forty nights. (Genesis 7:11-12)

Biblical literalist, John Murray, in 1840, had used the scripture above to argue subterranean waters were a major contributor to the flood of Noah.[351] Brown's Hydroplate Theory provided a greater comprehensive explanation for the source of water for the "fountains of the great deep." His Hydroplate Theory stated during the period before the flood, the waters were from subterranean reservoirs that (1.) had an average thickness of 3/4th of a mile; (2.) were located approximately ten miles beneath the Earth; (3.) were throughout the world in interconnected chambers; and (4.) had existed under high pressure and great heat as supercritical water.[352]

Rupture Phase of the Hydroplate Theory. Brown claimed the flood began when a microscopic crack occurred in the stretched crust and spread ten miles down to one of the subterranean chambers of supercritical water. This caused a rupture that raced around the world within two hours. The water roared out of the expanding rupture like a supersonic fountain. Some of it fell back as torrential rain. The more energetic aqueous fluids jetted above the atmosphere at supersonic speeds to fall back as frozen masses of mud. The great hailstorm flattened small animals and fractured the bones of large creatures such as the Berezovka Mammoth.[353] Brown indicated some debris from the very energetic regions of the fountains exceeded escape velocity to become comets,[354] asteroids, or meteoroids.[355]

Flood Phase of the Hydroplate Theory. The fountains of water caused strong rain storms ("…the windows of heaven were opened"). Supercritical waters from the subterranean chambers contained dissolved minerals. As the flood swept across the Earth, the minerals precipitated out of the water to form limestone deposits and great salt domes. Sediments from the flood waters settled across the Earth to commence the fossilized records of buried and trapped animals and plants.

Continental-Drift Phase of the Hydroplate Theory. As the rupture widened, the basalt rock beneath the exposed chambers sprung upward, forming the Mid-Atlantic Ridge.[356] The cascade of buckling basalt beneath the rupture shoved the continental plates away. Those plates had moved rapidly because the remaining subterranean waters served as lubricants. Pre-flood mountains were carried along and distorted. As the plates collided and/or grinded to a halt, they suffered compressions and buckled to form new mountains and valleys. Near the end of the Continental-Drift Phase, some of the buried vegetation and animal were caught up in the compressions that generate heat and pressure that started the formation of coal and oil deposits.[357]

Hunter's examination of Genesis, structured Noah's flood into three stages. The water level rose during the first forty days. Stage two was characterized as a period of steady water levels from the 40th through the 150th day. And, the third stage was the declining water level from the 150th to the 371st day.[358] The Recovery Phase of Brown's theory corresponds to Hunter's stage three of the flood. Water levels dropped as the buckling plates altered and created chambers for the subterranean waters.[359]

The Recovery Phase of Brown's suggested his first prediction in 1980. Prediction One stated: "Beneath major mountains are large volumes of pooled salt water."[360] The *Holy Bible* contains descriptions of subterranean waters that support Prediction One. Consider the following:

> Thou shalt not make unto thee any graven image, or any likeness of any thing that is in heaven above, or that is in the earth beneath, or that is in the water under the earth. (Exodus 20:4)

The above commandment indicates that none of the things within subterranean waters are to be worshipped. This lets us know subterranean aqueous fluids do exist. A popular Sunday School story contains a more explicit reference to water beneath mountains:

> I went down to the bottoms of the mountains; the earth with her bars was about me for ever: yet hast thou brought up my life from corruption, O Lord my God. (Jonah 2:6)

Jonah stated the great fish, that had been prepared to swallow him, had swum beneath mountains to deliver him to the capitol of the

Assyrian Empire within three days and nights (Jonah 1:17). It would have taken the animal weeks to swim through the Mediterranean Sea, the Gulf of Suez, the Red Sea, around the southern tip of Arabia, north in the Arabian Sea, through the Persian Gulf, north in the Euphrates River, and then north in the Tigris River to deposit him on the shores of Nineveh. The surface water route from Joppa around Arabia into the Tigris River to Nineveh would have covered over 4,700 miles. Such a journey would have taken over a week, but Jesus Christ stated it took only three days:

> For as Jonah was three days and three nights in the belly of the whale, so will the Son of man be three days and three nights in the heart of the earth. (Matthew 12:40)

According to Jonah 2:6, the great fish used a subterranean watercourse to make the trip. The shortest distance between Joppa and Nineveh is approximately 500 miles over land. A mountain range flanking the northern segment of the Fertile Crescent is between the two cities. A journey over that shorter distance would have taken much less time. To travel 500 miles in three day requires an average speed of only 6.9 miles per hour.

It is unlikely God had prepared a whale to transport Jonah. Whales can hold their breath for only an hour to swim beneath the waves. To swim the subterranean watercourse to Nineveh would have taken six days if the Jonah's fish swam at the speed of a Hawaiian humpback whale.[361] Such a mammal could not hold its breath long enough to complete the journey.

Jonah's fish may have been from the shark family. Great white sharks are large enough to swallow a man and do not have to hold their breath for oxygen. Unfortunately, their speed is much slower than whales. The great blue shark has the speed to make the journey within eleven days.[362] If the prepared fish had been an enlarged great blue shark, it could have delivered Jonah. Swift undersea currents could have swept them along to complete the trip within three days.

Jonah may have died and experienced the equivalent of an out-of-body experience. His normal senses could not have distinguished between swimming beneath waves or mountains. He probably expired before the great fish had entered the subterranean passage. God had revealed to Jonah his journey had been through a waterway beneath mountains.

The phenomena at Mount Sinai may be an explicit example of electromagnetically induce currents producing an unusually long earthquake:

> On the morning of the third day there were thunders and lightnings, and a thick cloud upon the mountain, and a very loud trumpet blast, so that all the people who were in the camp trembled. Then Moses brought the people out of the camp to meet God; and they took their stand at the foot of the mountain. And Mount Sinai was wrapped in smoke, because the Lord descended upon it in fire; and the smoke of it went up like the smoke of a kiln, and the whole mountain quaked greatly. (Exodus 19:16-18)

If the Lord's fire had been a large ball of plasma and if Mount Sinai was above pools of water, the quaking may have been caused by the electromagnetic heating of that water. Lightning is one of the products of intense electrical activity. Plasma is ionized gas that responds to and induces electromagnetic fields. Several attempts to explain ball lightning incorporate various plasma physics theories. The description above, about the phenomena at Mount Sinai, may be another example of electrically induced quaking. Normal earthquakes are brief. Mount Sinai quaked when the fire descended upon it and continued to shake as God verbally stated the Ten Commandments. The unusual seismic activity was probably caused by the concussive turbulence of electromagnetically heated water beneath the mountain.

A question that frequently arises around these works is "How can worldwide chambers of water be located ten miles within the Earth?" The pressures and heat would have prevented the seepage of water down to the mantel. The answer is that it was located there originally during the creation process. Consider the following verses:

> And God said, Let the waters under the Heaven be gathered together unto one place, and let the dry land appear: and it was so. And God called the dry land

Earth; and the gathering together of the waters called he Seas: and God saw that it was good. (Genesis 1:9-10)

For this they willingly are ignorant of, that by the word of God the heavens were of old, and the earth standing out of the water and in the water: (2 Peter 3:5)

God had created water first. He then separated them to create vast gaps between balls of water. One of the balls was Earth. Then, by transmutation, he converted regions of the balls into various forms of solid matter. The worldwide, interconnected subterranean chambers of aqueous fluids had been one of the features created in the Earth. Humphreys has used magnetic moments and the age of creation (6,000 years) to argue the Sun, Earth, Moon, and planets had been transmuted from water.[363]

Section 6-4. Evidence of Water Beneath Mountains and Islands

The previous section presented biblical scriptures and bible-based theories that support Prediction One of the Hydroplate Theory. This section lists the geophysical findings that evince Prediction One in terms of three different types of witnesses: magnetotelluric, seismic, and laboratory simulations.

Seven years after Brown had announced Prediction One, Hutton, Gough, Dawes, and Travassos detected a high conductivity region beneath the Canadian Rocky Mountains that exhibited the characteristics of hot saline water of mantel origin.[364] In 1991, deep water was proposed as an explanation for a class of earthquakes.[365] Brown, in 2001, stated a one-mile-thick layer of Brine salt water had been detected 10 miles below the Tibetan Plateau. His assertion was based on the magnetotelluric studies conducted by Project INDEPTH (International Deep Profiling of Tibet and the Himalaya). INDEPTH scientists had evinced a layer of aqueous fluid beneath the world's largest mountain range, the Tibetan Plateau.[366] The high conductivity was attributed to partial melt and aqueous fluids. Further magnetotelluric readings and analyses by that team yielded a model with a thin aqueous layer overlying a thick partial melt.[367] Magnetotelluric studies conducted by Patro, Brasse, Sarma, and Harinarayana of the crust below the Deccan Flood Basalts, India, detected subtrappean-hidden faults that were enriched with fluids.[368] Electrical resistance anomalies beneath central Taiwan were explained in terms of water and partial melts in the Earth's crust.[369]

Seismological studies evinced deep water. Helffrich and Wood attributed seismic contradictions to mantle water content.[370] Van der Meijde, Marone, van der Lee, and Giardini have acquired seismic evidence of water beneath the upper mantle transition zones of the Mediterranean.[371] During the following year, their analyses of seismic waves from beneath the Mediterranean Sea yielded evidence for the presence of deep water inside the upper mantle.[372]

Various laboratory simulations of heat, pressure, and chemical properties of minerals have suggested the presence of water in the Earth's mantle. Meade and Jeanloz used laboratory results to propose dehydration and high-pressure instabilities as a mechanism for earthquakes occurring at depths between 100-650 kilometers.[373] Green and Green promoted dehydrated embrittlement as the sole, experimentally confirmed mechanism that can cause faulting at depths from 50-300 km of a subducting lithosphere.[374] These findings evinced the existence of water beneath the mountains and offer a faulting mechanism.

Brown's theory has been comprehensive and controversial. The Hydroplate Theory supported the biblical account of the flood of Noah in every detail. It also indicated how the violence of the flood created the Grand Canyon, Mid-Oceanic Ridge, Continental Shelves and Slopes, Ocean Trenches, Seamounts and Tablemounts, Earthquakes, Magnetic Variations on the Ocean Floor, Submarine Canyons, Coal and Oil Formations, Methane Hydrates, Ice Age, Frozen Mammoths, Major Mountain Ranges, Overthrusts, Volcanoes and Lava, Geothermal Heat, Strata and Layered Fossils, Limestone, Metamorphic Rock, Plateaus, Salt Domes, Jigsaw Fit of the Continents, Changing Axis Tilt, Comets, and Asteroids and Meteoroids. Prediction One was confirmed by magnetotelluric, seismological, and chemical evidence for deep, aqueous fluids. But, his theory has not been regarded by the scientific community to be founded on science. Many creationist organizations such as Answers in Genesis and the Institute for Creation Research have considered his theory to be unworkable. On the other hand, over forty-three million viewers in America and Canada in 1993 saw a five-minute animation of the Hydroplate Theory, narrated by Dr. Brown, in the two-hour broadcast by CBS about Noah's Ark. The response to the Hydroplate Theory presentation of the program was very positive.

It is not preposterous to assume the dislocation of mountains and islands will be caused by the turbulence of aqueous fluids responding to geomagnetic currents that will be induced by the storms on the blackened Sun. Shaltout, Tadros, and Mesiha have shown solar activity can influence earth seismicity.[375] The mechanism for shifting the location of mountains and islands can be driven by the magnetic storms of the Sun caused by the opening of the Sixth Seal.

Chapter 7
The Great Migrations

> And the kings of the earth, and the great men, and the rich men, and the chief captains, and the mighty men, and every bondman, and every free man, hid themselves in the dens and in the rocks of the mountains; (Revelation 6:15)

This passage described mankind's response to the events stemming from the opening of the Sixth Seal. A literal interpretation implies the entire population of the world successfully finds shelter. A lottery system will not be used to select individuals to be saved from the catastrophes described in the remaining chapters of the Book of Revelation – none will be left outside.

Section 7-1. All Will be Able to Hide

Revelation 6:15 list the civil categories of mankind who will be sheltered. The phrase "kings of the earth" has been interpreted to mean both literal kings and the leaders of people such as heads of states, presidents, Emirs, and sheiks. The Greek word [SEC 3175] that was translated into "great men" is equivalent to the nobles, magnates, tycoons, and industrialists. The "rich men" are simply the wealthy and "chief captains" [SEC 5506] are military commanders with authority over at least a thousand troops. One may be tempted to assume that the word "every" denotes each type of bondman (slave) and freeman and not every individual who belongs to that category. Other translations (e.g., the *Living* Bible, *Simple English*, and, the *New Jerusalem with Apocrypha*) clearly indicate all hid themselves. The order of the description may indicate the last group to reach the shelters will be freemen and some of the first will be kings.

An astounding feature of this particular scripture is the use of the word "every." It was not used to describe the leaders and wealthy in this passage. The explanation for this may be a noble one. Some of the heads of state, pillars of society, and wealthy may have stayed behind to assure the complete evacuation of their regions.

The classifications in this passage do not refer to groupings by nationality, ethnicity, and/or language. For example, consider the following:

> And they of the people and kindreds and tongues and nations shall see their dead bodies three days and an half, and shall not suffer their dead bodies to be put in graves. (Revelation 11:9)

The absence of sociological groups may indicate the shelters will be filled with people from different countries, races, and languages. Civil management of the (1.) evacuations from cities, (2.) migrations across country to the shelters, (3.) admittance to shelters, and (4.) governance of the underground populations will be a horrific task. Cellular phones may be the principle instrument used to guide populations to their shelters during that period.

Previous evacuations shall pale in comparison with this global migration to underground facilities. History is replete with the accounts of thousands of refugees that have fled from their war torn cities. People hid within sewers and subways for hours to escape the dangers of bombing raids during World War II. Less than two decades later, civil defense authorities constructed bomb and fallout shelters to protect populations from the nuclear detonations and radiation, respectively, of an anticipated third world war. But, tens of thousands times those numbers shall be forced to flee by the Sixth Seal events. Various census assessments for the year 2000 had estimated the world population to have exceeded six billion (6,000,000,000).

Verse fifteen indicates men will hide in the dens and rocks of mountains. Another meaning for the word that was translated from Greek to "dens" [SEC 4693] is hiding place or resort. It has been translated as den five times and once as cave in the *Holy Bible*.

A distinction can be made between dens and caves. The word "dens" appears with the word "caves" in the following passage:

> (Of whom the world was not worthy:) they wandered in deserts, and in mountains, and in dens and caves of the earth (Hebrews 11:38).

This passage indicates the two words are different. Also, the Greek for "caves" [SEC 3692] has a stronger distinction for a space in the ground that has been naturally formed.

Dens convey the meaning of facilities and/or rooms equipped to serve as comfortable, temporary dwellings. Such dens would be stocked with food, water, lighting, heat, power supplies, and comforts such as furniture and entertainment. Dens can not be formed at the last moment. Mankind will not have time to rapidly construct the numerous hiding places to accommodate world population during the opening of the Sixth Seal.

Undeveloped caves are inhospitable. Dr. Tony Walthom, a famous authority on caves, gave the following description: "There is nothing on the surface of the Earth which is at all similar to this world below."[376] Those natural openings in the earth lack water for sanitation, drinking, and human waste. People would have to bring their own sources of power for lighting, heating, air circulation, sanitation, and pesticides to hide in caves. The latter is for the variety of life that occupies caves. Troglobites (e.g., eyeless fish, beetles, various insects) cannot survive beyond the caves and trogloxenes (e.g., rats, various types of bats) use caves for hibernation, sleep, and/or procreation.[377] Therefore, the majority of the 14,556 known caves in the Alabama, California, Indiana, Kentucky, Missouri, Oregon, Virginia, and Washington[378] could not serve as long term shelters. This situation would apply to many other regions throughout the world.

The Greek word [SEC 4074] for "rock" has been used for large stones, cliffs, and crags. Hiding places have been cut from large rocks within mountains. But, cliffs serving as adequate hiding places during the Sixth Seal events indicate severe weather, gas, and/or disease would not be the mechanisms that drive people into hiding. Cliffs provide protection from falling threats such as rain, hail, cosmic rays, and ultraviolet light. Unusually energetic solar cosmic rays seem to be the best candidate for the Sixth Seal mechanism that drives populations to the cliffs and underground facilities. Cliffs would be more suitable for those who suffer from claustrophobia.

Another interpretation of the Greek word [SEC 4074] for "rock" applies to mountains. A survey of literature about underground facilities by Kao contained further clarification of this scripture:

> Many of the installations are built either underground or in the sides of mountains. Many of the installations are tunneled into rock in the mountainsides which is relatively fault-free and is not prone to flooding during construction. Often, the rock is so strong that the tunnel walls do not have to be lined.[379]

An example of such a facility is the North American Aerospace Defense Command (NORAD) located within Cheyenne Mountain, Colorado, and the underground headquarters of the Strategic Air Command, Offutt Air Force Base, Omaha, Nebraska. These facilities have been constructed to withstand several direct hits by nuclear warheads. Most underground structures would not be that strong. Most of the world populations may be protected by the equivalent of fallout shelters equipped with thick shielding.

Cliffs will be one form of shelter they will occupy. Some have been large enough to accommodate hundreds. For example, the Anasazi Indian homes nested beneath the overhanging cliffs of Mesa Verde, Colorado.[380] Tourists go there to see the houses and streets that were built into the cliffs. The Anasazi abandoned their dwellings during the fourteenth century.

Europe has several nations that have constructed mountain shelters. If the global evacuations commenced today, Switzerland would be able shelter its entire population. Its underground shelters can accommodate seven million people[381] and 85 percent of all its citizens could reside in blast shelters.[382] The region of nations in northwestern Europe known as Scandinavia may approximate Switzerland's achievement. According to Kao: "The Scandinavian countries have built many underground or mountainside structures for civil defense. The mountainous terrain provides a very hardened personnel shelter compared to what can be built aboveground."[383] He also stated: "Scandinavian countries have been using rock caverns extensively and have developed design and construction experience."[384] Deep rock caverns can provide large underground spaces.

Many hiding places will be available at the beginning of the Sixth Seal events. Tunneling technology has produced thousands of underground facilities since the time of Apostle John's vision. During his era, slave labor used chisels, hammers, and pickaxes to dig mines and tunnels. The invention of gunpowder enabled miners to drill and blast around the thirteenth century. Holes for the explosives were drilled into the rock. The explosives were inserted and detonated. Laborers shoveled away the debris and the process was repeated. Engineering breakthroughs during the early 1950's created tunnel boring machines (TBM). The TBM accelerated the construction of underground facilities during the 1960s and 1970s.[385]

There are numerous manmade buried facilities. Richard Sauder authored two popular books that list several underground[386] and underwater[387] bases and tunnels. His works have evinced the multitude of underground facilities (UGF) that have been constructed. For example, his latter work contains a lengthy list of tunnels that have been constructed in the western states by the U.S. Bureau of Reclamation.[388] Excellent summaries on the changes in excavation technology that allowed the accelerated growth of tunneling projects have been produced.[389, 390]

Numerous regulations and guidelines for the planning and management of UGF projects have been published. The body of United States regulations and standards, stemming from the Occupational Safety and Health Act, has been promulgated for the underground construction industry.[391] Tarkoy had presented a series of guiding principles that tunneling contractors could use for the preparation of claims.[392] Oggeri and Ova[393] and Tarkoy and Wagner[394] have produced guidelines for assuring tunneling quality. A search of the Internet yields numerous articles and web sites about UGF and TBM.

Numerous dens will be available throughout the globe to accommodate the world populations fleeing from the Sixth Seal events. Those facilities will serve as a refuge from subsequent catastrophic judgments that will transpire during the Day of Wrath (e.g., the meteoroid impact of Revelation 8:8). Their numbers will be large enough to provide shelter for the entire population of the world. That would be six billion people if it were to occur now.

Section 7-2. Creating Enough Shelters before the Sixth Seal Events

Conditions prior to the Sixth Seal events may have caused each nation to produced effective emergency management agencies. Each would have developed and acquired experience with the disaster mitigating factors of preparedness, response, and recovery.[395]

The troublesome times prior to the Sixth Seal events may cause countries to develop effective disaster mitigating systems. A period known as the "beginning of sorrows" precedes the opening of the Sixth Seal. The Beginning of Sorrows shares many of the characteristics associated with the four horsemen of the Apocalypse. Descriptions of it by Jesus Christ had been recorded in Matthew 24:4-8; Mark 13:8; and Luke 21:8-11. "Beginning of sorrows" is an Old English phrase that was translated from a Greek word [SEC 5604] that means birth pains. As pregnant women approach delivery, their pains from contractions both intensify and occur with increasing frequency. Some of the incidents Jesus cited for that period are wars between ethnic groups, wars between kingdoms, and plagues. They could stimulate evacuation and disaster management systems development. And, the fourth horseman of the apocalypse, Death, causes the death of one fourth of the world population (Revelation 6:8). It is not clear if his task is completed before the opening of the Six Seal.

Natural hazards from space may be the incentive for world wide shelter construction projects. Various studies have indicated a risk does exist of a nearby star exploding into a supernova (e.g., WR104). Radiation levels in the vicinity of the Earth could increase a thousand times over normal. Though the probabilities of such events are low, the technology currently exists to allow the detection of early warning signs. Forecasts based on the data from muon telescopes and neutron monitors could give Earth's civilization a few decades to develop alternative life underground for hundreds of years.[396] Such concerns over the threat of a supernova explosion may cause mankind to build enough underground shelters to accommodate all of the populations of the world before the opening of the Sixth Seal.

All Shall Hide

UGF have been constructed for several reasons. The energy costs for maintaining underground bases can be smaller than corresponding expenses for conventional, above ground buildings over a twenty- to thirty-year life cycle.[397] Sterling and Godard[398] cited the following reasons that have been used for developing underground facilities: "(3.1) land use and location reasons," "(3.2) isolation considerations," "(3.3) environmental protection," "(3.4) topographic reasons," and "(3.5) social benefits." Photographic examples of UGFs for reasons 3.1 through 3.5 have been made available by the International Tunneling Association – Association Internationale des Travaux En Souterrain.

Many of the facilities stemming from reason component "(3.2.4) containment" will not be able to serve as shelters during Revelation 6:15. Many of the UGF from this category process and/or store various forms of hazardous materials.

The rationale for creating most of the deeply buried military facilities had stemmed from the "(3.2.3) protection" and "(3.2.5) security" components of (3.2). There are many examples from history that illustrate the protective ability of UGF. The underground production facilities of the German V-2 rockets at Nordhausen, south of the Harz Mountains, were not discovered until the close of World Ward II.[399]

The threat from weapons-of-mass-destruction has not been the sole incentive for numerous international UGF projects. Precision guided munitions of the First Gulf War demonstrated the vulnerability of above-ground buildings. Pictures of a laser-guided bomb flying through a window to destroy a building caused third world countries to commence UGF construction projects.[400, 401] Estimates from the intelligence community, cited in a July 2001 report to the Congress, sets the total, worldwide number of hard and deeply buried targets to be over ten thousand.[402] The recent construction of UGF has caused the promulgation of proposals[403] for the development of new nuclear earth penetrating weapon systems.

Locating military underground facilities is very difficult. History has shown the most effective tunnel detection tool has been interrogation. For example, two women revealed to General Patton's Third Army the location of a salt mine, over two thousand feet beneath the earth, that carried one hundred tons of gold bullion and millions in art and currency.[404] Wernher von Braun, the head of the team that developed the German V-2 rocket, had arranged the surrender to the American Army of his group and the transfer of three hundred boxcar loads of V-2 rocket materiel that had been hidden in an underground factory near Niedersachswerfen.[405] During the twenty-three years following 1974, only twenty-three tunnels, beneath the demilitarized zone in Korea, had been located by the United Nations Tunnel Negotiation Team.[406] The interrogations of relatives and former bodyguards of Saddam Husein led to his capture at his hole in the ground on Saturday, December 13, 2003.[407] Prisoner and refugee interrogation has been the most effective tool for locating the tunnels. The numbers of underground facilities will probably continue to increase after the second seal has been opened to release the second horseman of the apocalypse (Revelation 6:3-4).

The difficulties associated with the detection and characterization of UGF has generated research in areas of physics beyond the diverse applications of acoustics. Feasibility studies for remote UGF identification by the legendary High frequency Active Auroral Research Program (HAARP) Ionospheric Research Instrument have been conducted.[408] Several achievements may be attained for detection but, electromagnetic fields and sound waves can be transmitted to offset attempts to remotely identify UGFs. Gravity seems to be the sole force field that can not be manipulated. Subsequently, research into gravimeters and gravitational wave generators for UGF identification has been conducted.[409] Breakthroughs in these areas, for military applications, seem to be many years into the future. Subsequently, third world countries will continue to develop subterranean facilities that escape detection by more technologically advanced nations.

Scientific curiosity in certain areas of physics and biology has driven projects for deep underground laboratories. For example, the world's next-generation of particle accelerator, the Large Hadron Collider, is located deep beneath the Swiss-French border.[410] The February 2010 issue of *Symmetry*, a joint Fermilab/SLAC Publication, was dedicated to deep underground laboratories.[411] It included very readable reviews of the Laboratori Nazionali del Gran Sasso, Italy, the largest operating underground science laboratory in the world and the Sudbary Neutrino Observatory Laboratory, Ontario, Canada, the deepest operating laboratory in the world. And, it contained summaries about the proposals to develop the Homestake Mine, Lead, South Dakota, and the Chin

JinPing Deep Underground Laboratory, Sichuan Province as contenders for the future world titles of both the deepest and largest laboratories.

The financial hardships that erupted during the early part of the twenty-first century have not caused the cancellation of all underground facility construction projects. Many of them shall be directed by world leaders to open their doors to the world populations after Jesus Christ has opened the Sixth Seal.

Figure 5. Tunnel boring machines, like the one above, have enabled contractors to build underground facilities more rapidly than by the earlier methods of blasting and shoveling.

Taylor A. Cisco, Jr.

Section 7-3. Penetrating Earth's Shields

What mechanism will drive people to seek shelters? The examination of the Greek word [SEC 4074] for "rock" in Section 7-1 suggested the threat emanates from above. Cliffs could not provide adequate protection from heat, cold, poisonous gas, and/or floods. The threat has to be more penetrating than rain and ultraviolet light. Tents and layers of cardboard would have been sufficient if those had been the threat. And, as discussed in Section 4-1, radial points, the principal feature of meteor showers, disqualifies them as the threat. Intensified particle radiation from the blacken Sun seems to be the answer.

The Earth's magnetic field is the first line of defense. Note the statement from the NASA Goddard Spaceflight Center:

> Behind the light show lies enough flux of energetic particles carried by solar wind to render our planet uninhabitable. The Earth's magnetic field, also known as the magnetosphere, is the only thing that shields us from the Sun. Even the magnetosphere cannot fully guard us from the wrath of the Sun.[412]

The second line of defense is the Earth's atmosphere. It serves as a thick layer of protection from solar radiation. With respect to its shielding effects, it is equivalent to twelve feet of concrete.[413]

Is their evidence of space weather phenomena causing increases in radiation at the surface? The answer is yes. On rare occasions, high energy solar proton events (SPE) have penetrated Earth's magnetic field and its atmospheric shielding effects. An event in which the Sun has produced particles of sufficient energy and intensity to increase radiation levels on the surface of the Earth is known as a Ground Level Enhancement (GLE). Seventy GLEs have been observed from February 28, 1942, through December 13, 2006. The August 1972 SPE that alarmed Apollo Moon Program managers (see description in Section 5-3), had generated GLE24 and GLE25. GLE42 exemplified the ability of energetic particles to penetrate hundreds of feet of rock and soil. The energies and intensities of muon secondary cosmic rays of GLE42, that had penetrated 230 meters (approximately 690 feet) to the base of Mount Andyrchi on September 29, 1989, were measured

All Shall Hide

by the Bakson Underground Scintillation Telescope.[414] The CME known as the "Mother of all Halos" had caused GLE65 on October 28, 2003 that was later called the "Greek Effect."[415] A narrow stream of relativistic neutrons caused GLE68 of January 17, 2005.[416] GLE70 of December 13, 2006 was another example of intense regional effects. Only the neutron monitors of the LARC station (King George Island, Antarctica) registered GLE70's surge in cosmic ray intensity.[417] The other three stations in the IFSI-INAF partnership did not record any significant changes. This set of examples was selected to illustrate both the global and regional characteristics of GLEs. More severe GLEs will be produced by the solar blackening of the Sixth Seal events.

Section 7-4. Aerospace Hazards

Several studies have recognized the risks of lethal radiation that can be encountered by astronauts. Health hazards caused by solar storms have been the subject of research for several years. Reames has identified the types, energies, and timing of particles that pose a normal threat.[418] The focus of such studies has been the ability of ionizing radiation to damage organ tissue.

Populations in high-altitude cities receive an average of seventy percent more annual exposure to neutron cosmic rays than commercial aircraft operations.[419] Flight crews are exposed to higher levels of radiation during their relatively brief flights. But, the total exposure that has been accumulated for a year for citizens of high altitude cities is greater. La Paz, Bolivia; Lhasa, China; Quito, Ecuador; Mexico City, Mexico; Nairobi, Kenya; Denver, United States; and Tehran, Iran are among the list of cities in the high-altitude category. Biological response data for such exposures to high-energy neutrons has been deemed to be inadequate. This section reviews the radiation hazards that astronauts and flight crews face and the next section examines recent findings for ground level risks.

Military pilots face lower risks of exposure to radiation because their accumulated flight times are smaller than those of commercial aircrews. The responses of the blood forming organ and selection criteria for effective shielding for astronauts have been identified.[420] Normal radiation levels experienced by the airlines have been the subject of considerable research. Several studies have been conducted on the risks of increased incidence of various cancers aircrews face from exposure to cosmic radiation.[421] The focus of research included breast cancer,[422] skin cancers (melanoma, non-melanoma, and basal carcinoma),[423] and prostate cancer. Rates of skin cancers for aircrews and pilots seem to match the general population rates while breast cancer and prostate cancer incidents have merited additional research.[424] In general, the risks have been deemed to be tolerable with monitoring and licensed activity. The Nuclear Regulatory Commission promulgates standards for protection against radiation in the Federal Register (e.g., 63 FR 39477-39483).

The National Oceanic and Atmospheric Administration (NOAA) provides alerts whenever solar events pose a hazard to satellites, astronauts, aircrews, and power stations.[425] The codes for indicators of radiation storm severity range from S1, for minor, to S5, for severe. The S5 code corresponds to particles with 10 MeV or more of energy with fluences reaching or exceeding 10^5 particles per cm^2 per steradian per second. Such dosage levels are equivalent to 100 chest X-rays. The expected occurrence rate of S5 radiation storms is one day per solar cycle (11.4 years).

Radiation levels will have to be dramatically increased to cause the fall of aircraft due to flight crews suffering from radiation sickness. The magnitude of those radiation storms would not limit the catastrophe to just the Sun-side of the Earth. Solar energetic particles are guided by magnetic lines of force. Radiation would follow those lines, in varying intensities, throughout the latitudes of the globe. The falling lights of aircraft would be seen around the world.

Dosages of the Sixth Seal radiation storms could exceed 5,000 rads. Immediate incapacitation and death within a week are the human health effects. Heavy shielding against galactic cosmic rays may protect the life of astronautics and the functionality of electronic circuits within satellites. But, air traffic would suffer losses.

Acute Radiation Syndrome (ARS) (also known as radiation sickness and radiation toxicity) occurs if the radiation dose was larger than 0.7 Gray (70 rads); was from an external source; was caused by high energy X-rays, gamma rays, and/or neutrons; radiated the entire human body; and exposure had occurred over a few minutes. The three primary syndromes of ARS are bone marrow, gastrointestinal (GI), and cardiovascular (CV)/central nervous system (CNS) syndromes. Death from infection and hemorrhaging, caused by the destruction of bone marrow, is the result of 0.7 to 10 Grays (70 – 1,000 rads). Death within two weeks, due to the GI syndrome of ARS, can be caused by 10 to 100 Grays (1,000 – 10,000 rads) of radiation. Death from the GI syndrome stems from damage to the GI that will yield dehydration and electrolyte imbalance. Death within three days, due to the CV/CNS syndrome of ARS, can be caused by dosages equal to or greater than 50 Grays (5,000 rads). Death by the CV/CNS syndrome stems from the collapse of the circulatory system and increased pressure within the skull due to edema, vasculitis, and meningitis. The four stages of ARS are the Prodomal Statge (nausea, vomiting, and nausea); Latent Stage (a momentary period from a few hours to weeks where the victim feels well); Manifested Illness Stage (emergence of the symptoms from specific ARS syndromes); and the Recovery or Death Stage - which can span between a few weeks to two years.[426]

The kill mechanism that causes aircraft to fall from the sky and populations to flee to underground shelters does not seem to be ARS. The tone of Revelation 6:12, 15 implies a sudden onslaught. For instantaneous incapacitation under ARS, radiation levels would have to be much larger than 50 Grays (5,000 rads). It is hard to imagine sudden radiation levels exceeding 5,000 rads occurring within the altitudes of aircraft. Such levels are tantamount to the effects caused by neutron-bomb explosions. Other mechanisms have been observed at ground level, with much smaller radiation dosages, that have yielded sudden deaths. The next section examines other fatal effects that are more suitable explanations for Revelation 6:13, 15.

Section 7-5. Ground Level Hazards

The focus of the previous section examined how ionizing radiation could damage organ tissue to an extent resulting in death. This sections reviews fatal health problems that can be caused by geomagnetic storms and cosmic rays. Using a very general analogy, studying how radiation damages organs, in the previous section, was similar to studying how machine guns can bring down airplanes. Correspondingly, the reviews in this section are similar to studies on how weak signals from cell-phones and lap-tops can bring down aircraft. Just as faint signals from such items can be misinterpreted by the aircraft's electronics as commands from the flight crew, GMA and CRA can cause death in a segment of the human population.

Research has shown the human body is sensitive to variations in space weather (e.g., blood pressure changes with geomagnetic activity[427]). Scientists have referred to this field of study as heliobiology (also referred to as cosmobiology, astrobiology, and clinical cosmobiology[428]). Heliobiology examines the responses of human biological systems (circulatory (cardiovascular), digestive, endocannabinoid, endocrine, integumentary, lymphatic, muscular, nervous, reproductive, respiratory, skeletal, and urinary) to variations in solar, geomagnetic, and cosmic ray activity.

People will become more fearful than normal during the opening of the Sixth Seal due to the geomagnetic activity induced by the blacken Sun. A prophecy recorded by Luke provides a connection between the fears that will sweep across mankind and the astrophysical disturbances:

> And there shall be signs in the sun, and in the moon, and in the stars; and upon the earth distress of nations, with perplexity; the sea and the waves roaring; Men's hearts failing them for fear, and for looking after those things which are coming on the earth: for the powers of heaven shall be shaken. (Luke 21:25-26)

That scripture lets us know the induced fears will be strong enough to cause heart failure. Some studies have shown that negative emotional backgrounds[429] and traffic accidents[430] have increased with intensified geomagnetic activity.[431] One of the newspaper stories selected to illustrate the impact of the Carrington Event was about a terrified sixteen year old girl, from Columbus, Ohio.[432] The October 8, 1859 *Harpers Weekly* article stated the Sheriff of Ottawa County had to take her into custody and she was later placed in a lunatic asylum. The article's reporter indicated it had transpired during a time when "… fears of the end of the Earth were making their rounds in many religious circles." It should be noted that several historical documents summarized earlier

All Shall Hide

in this book as evidence for ASDEs contained accounts of terrorized citizens. See Section 2-5 for the descriptions of hysteria in Kerkade, Netherlands, A.D. 1133; Lucca, Italy, and Split, Croatia, A.D. 1239; and Split, Croatia, A.D. 1241.

It should be noted that the Luke 21:25-26 also refers to the "seas and waves roaring." The magnetic fields conveyed to the Earth by the huge ejections of plasma from the Sun have caused it to shake very slightly.[433] Intensified heliomagnetic activity associated with the blacken Sun may be strong enough to cause the Earth to wobble along its orbit around the Sun. Such wiggles, too gradual to be felt as earthquakes, may resonate long enough to cause the waters to slosh around to yield roaring seas and waves.

Intensified geomagnetic activity may create the emotional state of fear during the opening of the Sixth Seal, but it is not the mechanism that causes people to flee for shelter. An examination of the Greek root for the word rocks in Revelation 6:15 indicates people will hide beneath cliffs. Such outcroppings can not protect them from fluctuations in the Earth's magnetic field. Cliffs, as shields, implies the threat falls to the ground like rain, debris, ultraviolet light, or cosmic rays.

Another argument against geomagnetic activity as the kill mechanism is its inverse relationship with the cardiovascular system. Cardiovascular problems occur more often during quiet periods of the Earth's magnetic field (Forbush decrease) than during high geomagnetic activity.[434] In 1999, Elyiahu Stoupel, Division of Cardiology, Rabin Medical Center, Petah Tiqwa, Israel, reported statistically significant inverse correlations between geomagnetic activity and cardiovascular parameters.[435] In 2006, Palmer's literature review concluded people located in the higher latitudes were more sensitive to geomagnetic disturbances; that cardiovascular health responded to extremes (high and low) in geomagnetic activity (GMA); between 10-15% of the populations studied were significantly affected by GMA; and heart rates varied inversely with GMA.[436] Therefore, the search for possible causes for flight to cliffs, caves, and dens can not stop with extreme geomagnetic activity.

Recent studies have shown deaths rates caused by diseases of the cardiovascular and the nervous systems have been aligned with changes in the cosmic ray neutron component of space weather.[437] Initially, such relationships were very controversial, but studies from Azerbaijan, Bulgaria, Greece, Israel, Lithuania, Russia, and Slovakia, have manifested statistically significant correlations between dysfunctional behavior of the cardiovascular system and variations in geomagnetic, solar, and cosmic ray neutron activity.

The primary function of the cardiovascular system is to circulate blood to the tissues and organs of the body. Scripture states the vital role of that system: "...For the life of the flesh [is] in the blood..." (Leviticus 17:11). Another term used for cardiovascular disease is heart disease. Each year, more Americans are killed by it than by cancer. The names of the various types of heart diseases are Coronary Artery Disease, Heart Attack, Irregular Heart Rhythm, Atrial Fibrillation, Heart Valve Disease, Sudden Cardiac Death, Congenital Heart Disease, Heart Muscle Disease, Dilated Cardiomyopathy, Hypertrophic Cardiomyopathy, Restrictive Cardiomyopathy, Pericarditis, Pericardial Effusion, and Marfan Syndrome. According to the literature, the types of diseases that have been found to be responsive to changes in space weather were acute myocardial infarction, sudden cardiac death (SCD), cardiac arrhythmia, brain stroke, and ischemic heart disease. Some of these terms do not seem to be included in the above list of types of heart diseases. But, a little research indicates most of those terms are components of particular types of heart disease. Acute myocardial infarction (AMI) is the medical term used for heart attack. Ischemic heart disease (IHD) develops as coronary artery disease progresses. Brain stroke is a disease of the nervous system, not the cardiovascular. It is caused by disrupted blood flow to the brain. Its medical term is cerebral vascular accidents (CVA).

Biochemistry,[438] congenital heart disease,[439] and immunological changes,[440] have been shown to be sensitive to space weather. But, medical disruptions like those are far less terrifying than experiencing palpitations and chess pressure while seeing others collapsing and dying during the blacken Sun of the Sixth Seal. Therefore, the emphasis and summaries presented in the remainder of this section shall focus on the responses of AMI, SCD, IHD, arrhythmia, and CVA to cosmic ray neutron component of space weather.

AMI, commonly known as a heart attack, is the interruption of blood supply to part of the heart, causing some heart cells to die.

The interruptions are frequently caused by clots blocking blood flow. Symptoms of a heart attack for men are chest pain spreading through the left arm and/or left side of the neck; difficulty breathing; palpitations; queasy stomach; vomiting; perspiring; and/or feelings of impending doom. Symptoms for women include difficulty breathing, weakness, and indigestion. Statistically significant correlations between AMI cases and cosmic ray neutron rates were manifested by Stoupel, et al in a study of 116,683 patients during 204 months in the Kaunas Registry of Lithuania.[441] A recent paper by Stoupel, et al examined the difference in sensitivities of AMI and intermediate coronary syndrome cases to the same concomitant cosmic ray neutron variations.[442]

SCD is used to describe the situation where the heart, without warning, suddenly stops pumping and causes the victim to immediately loose consciousness. Death normally follows within one hour. Over half of all deaths attributed to heart disease were caused by SCD. Certain malfunctions in the heart's electrical system cause it to stop. Emergency shocks generated by a defibrillator to the heart may restore its rhythm.[443] Unfortunately, with SCD, the heart resists attempts of resuscitation and early defibrillation.

The sudden death of young athletes attributed to SCD has drawn considerable media attention and subsequent resources for research. Those efforts have yielded methods for identifying the symptoms of those at risk; tests for predicting risks; and treatments for prevention. Babayev, et al reported direct relationships between increases in cosmic ray neutron rates and SCD mortality rates in Azerbaijan.[444]

IHD, ischemic heart disease, is characterized by ischaemia (reduced blood supply) to the heart muscle. Coronary artery disease frequently causes ischaemia. Aging, smoking, high cholesterol levels, diabetes, and/or high blood pressure increase the risk of IHD. It occurs more often in men and people with close relatives with IHD. Chest pain associated with physical exertion and weakened tolerances to exercise are the principle symptoms. Stoupel, et al reported correlations between IHD rates and cosmic ray neutron rates from a hundred ninety-two month study of 674,004 deaths in the Republic of Lithuania.[445]

Arrhythmia is a diversified set of conditions typified by various forms of high, low, and/or irregular heart beats that are caused by the body's abnormal electrical activity. Normal heart rates for adults range from 60 to 80 beats per minute and are much higher for children. A fatal cardiac arrest caused by an arrhythmia is called a sudden arrhythmia death syndrome (SADS). Gigolashvili, et al have reported periodicities in ventricular arrhythmias that were sensitive to the rotation period of the Sun.[446] Stoupel, et al reported rates of occurrence of ventricular tachycardial/fibrillation, recorded on implantable cardioverter defibrillators, had increased with the daily average rate of cosmic ray neutrons.[447] The December 4 through 24, 2006, study by Papailiou, et al of the influence of cosmic ray neutrons on the heart rates of patients of the KAT Hospital, Athens, Greece, included the effects of the December thirteenth GLE.[448] Mavromichalaki, et al, distilled statistically significant correlations between heart rates, GMA, and CRA from the collaboration of scientific groups from Athens, Greece; Kosice, Slovakia; and Sofia, Bulgaria.[449]

CVA, known as brain stroke, in lay terms, cerebral vascular accident in scientific literature, is the rapid spread of brain dysfunctions. It is frequently manifested by loss of the ability to move limbs and muscles on one side of the body; diminished ability to understand or formulate speech, or loss of a portion of eyesight on one side of the body. High blood pressure, aging, previous strokes or transient ischemic attack, diabetes, high cholesterol, smoking and atrial fibrillation increase the risk of CVA. The study of deaths in the Republic of Lithuania by Stoupel, et al also showed sensitivities of CVA rates with variations in cosmic ray neutron rates.[450]

Some research during the middle of the twentieth did not detect correlations between AMI and cosmic ray activity. For example, the application of the Correlation by Thirds statistical tool by Fogel and Righthand did not show any alignments between AMI and cosmic ray activity.[451] Their paper did not indicate which cosmic rays had been examined. More than half of the cosmic rays that reach the surface are muons – the remaining components are electrons, positrons, and photons. Neutrons, from the much rarer hadronic cascades also reach the ground. Fogel and Righthand did note the increase in sunspot activity had a moderate effect on all incidents of death except those caused by heart attack.

Several studies by Stoupel have shown direct variations between cosmic ray (neutron) activity (CRA) and sudden cardiac death. Stoupel reemphasized the need for additional research into examining the influence of cosmic ray neutrons.[452] In 2010, Stoupel, et al, showed that the indicators for acute myocardial infarction (AMI), and all of its subtypes, and intermediate coronary syndrome (ICS) change with variations in solar activity (SA), cosmic ray (neutron) activity, and geomagnetic activity (GMA).[453] The article was based on 72 months of data (2000-2005) from the MONICA International study in Kaunas, Lithuania, of 4,633 patients with AMI and 961 with ICS. It statistically manifested the direct variations of AMI with CRA; direct variations of ICS with GMA; and the inverse relationships of AMI with SA and/or GMA.

To understand cosmic ray neutron effects on the cardiovascular system, consider the 2006 article by Stoupel, et al on SCD data gathered continuously from January 2003 through December 2005 on the medical assessments of twenty-two Emergency and First Medical Aid Stations (EFMAS) spread throughout the mid-latitudes of the Absheron Peninsula, Azerbaijan.[454] Its capitol city, Baku, with over 3 million inhabitants, was included in the study. Stoupel indicated EFMAS had accurately registered all witnessed and un-witnessed SCD cases in accordance to ICD-10, 146.1 classification codes. The mortality rates caused by SCDs were compared with cosmic ray data from the Moscow Cosmic Ray Station (Neutron Monitor), IZMIRAN, Russian Academy of Sciences. The average daily neutron rate for the three-year period of 1,096 days was $8,475.35 \pm 339.7$ impulses per minute. A total of 788 SCDs occurred on 523 days of the three-year period and no SCDs occurred on the remaining 573 days. Their ages ranged from 25 to 88 for 666 males and 122 females. The average daily neutron rate for the 523 days of 788 SCDs was $8,538.08 \pm 322.5$ impulses per minute. Although the difference in average neutron rates was only 62.73 (0.74%), the 788 SCDs had occurred. One to five was the range of SCD cases occurring on the 523 days. Sixteen days had four to five SCDs and a corresponding average neutron rate of $8,657.5 \pm 189$ impulses per minute. The increase of 182.15 impulses per minute was equivalent to a 2.15% change. Stoupel's team showed the probability of the differences being attributable to random phenomena was only 0.0014 and its correlation coefficient was 0.511. The slight increases in the number of neutrons reaching the middle latitude of Baku, Azerbaijan, seemed to cause the increase in sudden cardiac deaths.

The above collection of studies suggests death rates during the events associated with the Carrington Flair should have been higher than normal. Marusek had studied some of the papers mentioned above and examined the Internet posted archives and death records of five Kentucky counties. He found the nineteen deaths that occurred during August 26-29; September 1-4, and October 10-13 in 1859 to be 1,986 times higher than the daily death rate for that year.[455] The geomagnetic storm and cosmic ray activity had accelerated the incidents of death among those who were older than fifty years of age. His findings were disseminated through his self-published whitepaper.

We see from the above studies that accelerated heart failure rates, not acute radiation sickness, will be the likely cause of terror stemming from the blacken Sun of the Sixth Seal event. If acute radiation sickness were the cause, we would expect radioactive contamination to be a problem for the survivors. The scriptures do not describe such phenomena. Interpreting Revelation 6:15 as physical phenomena is not preposterous. Increasing recognition by the scientific community of human sensitivities to space weather storms supports the deduction that heliophysically induced heart failures is a viable explanation for world populations migrating to shelters before the Day of Wrath of the Lamb.

Figure 6. Buildings that have not been damaged by the great earthquake of the Sixth Seal may provide enough shielding against cosmic ray neutrons to serve as effective shelters.

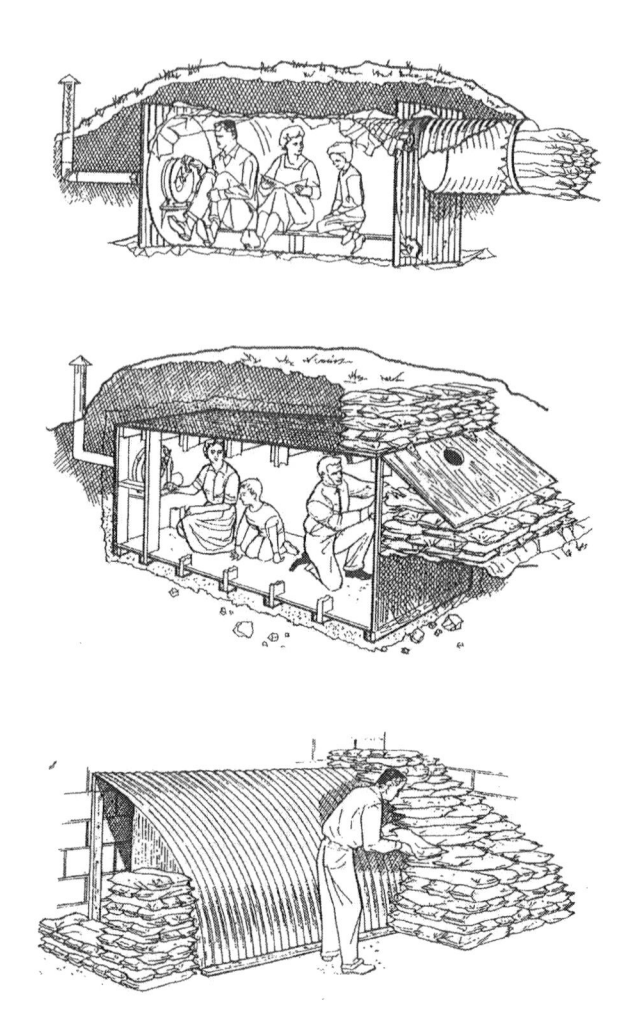

Figure 7. Private family shelters may not provide enough shielding against the Sixth Seal cosmic ray neutrons.

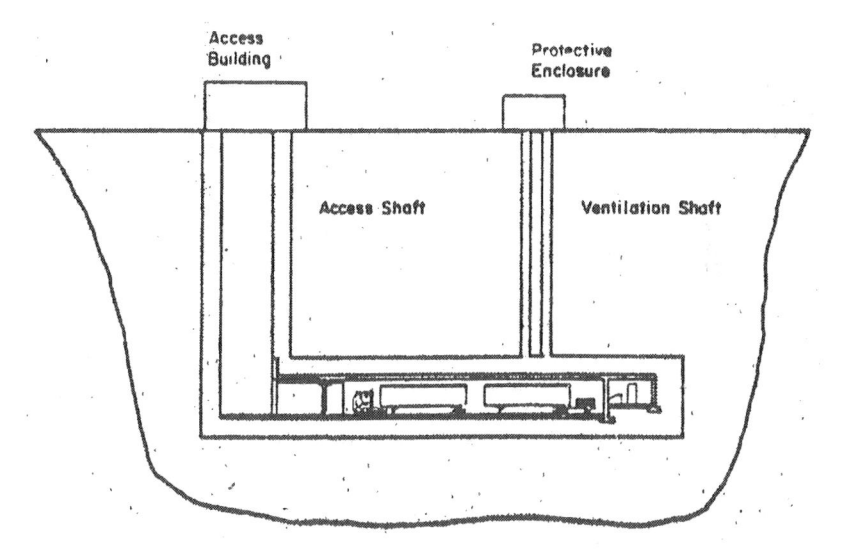

Figure 8. The majority of the world population will have to seek shelters with adequate shielding against the Six Seal cosmic ray neutrons.

Chapter 8
Correct Interpretation of Events

> And said to the mountains and rocks, Fall on us, and hide us from the face of him that sitteth on the throne, and from the wrath of the Lamb: For the great day of his wrath is come; and who shall be able to stand? (Revelation 6:16-17)

An examination of Revelation 6:16-17 indicates populations from different nations, ethnicities, and religions shall correctly understand the meaning of the Sixth Seal events. Nobody attributes them to Mother Nature. All of them will realize they have arrived at the dawn of the Lord's Day of Wrath. Inhabitants of the Earth may ask during the onslaught of those events: "Did the Rapture occur; has the Great Tribulation started; and/or has Christ returned?" But, everybody shall realize, near the close of the Sixth Seal events, that they have experienced the divine aggregate of terrifying signs that immediately precede the Day of Wrath. It shall be one of the very few moments in ecumenical history when all beliefs, faiths, and philosophies form the same deductions.

How will the sheltered survivors know the characteristics of the signs that precede the Day of Wrath of the Lamb? Jesus Christ provided an answer to that question during His Olivet discourse (Matthew 24:14): "And this gospel of the kingdom shall be preached in all the world for a witness unto all nations; and then shall the end come." The Gospel will have been preached throughout the world by the time the Sixth Seal is opened. Nations and tribes will not be able to claim they had not been warned.

The Sixth Seal events will cause extreme changes in life styles.

Mankind made several architectural achievements with the construction of skyscrapers. The advent of the Sixth Seal events shall drive mankind from luxurious spaces of lawns and bedrooms to cramped quarters beneath cliffs or within subterranean facilities. The blackened Sun drastically diminishes visible light for a period longer than the crucifixion. Daylight will not be taken for granted after the opening of the Sixth Seal.

The events following the opening of the Sixth Seal will be components of a pre-wrath sign that affects the solar system. It is a further elaboration of the prophecy recorded by Joel:

> And it shall come to pass afterward, that I will pour out my spirit upon all flesh; and your sons and your daughters shall prophesy, your old men shall dream dreams, your young men shall see visions: And also upon the servants and upon the handmaids in those days will I pour out my spirit. And I will shew wonders in the heavens and in the earth, blood, and fire, and pillars of smoke. The sun shall be turned into darkness, and the moon into blood, before the great and the terrible day of the Lord come. (Joel 2:28-31)

The locations of signs are important. They help people rapidly and accurately make decisions. The subject of those decisions may be financial, health related, and/or destinations. Highway exit ramp signs are not placed within buildings. People walking from one office to another do not need those signs. If that type of information is desired, they get it from maps. Ramp signs are placed next to highways. They are large and easy to understand. The first sign for a particular ramp may be placed several hundred miles a head of it. As the distance dwindles between drivers and an exit ramp, additional signs appear indicating the remaining miles and, finally, the remaining fractions of a mile or feet to the exit. The exit ramp signs have been constructed and located in a manner to allow drivers time to position themselves in the appropriate lane and to check for an adequate supply of gasoline. If they have only an eighth of a tank for an exit that is three hundred miles away, they will pull off of the highway to fill their gasoline tanks. Highway signs provide information to help drivers safely reach their destinations. Ramp signs within nurseries do not help expressway drivers. A physician does not measure tire pressure to determine the temperature of a patient. That information is obtained directly from the patient by devices such as thermometers, scanners, et cetera.

One of the earliest predictions of the pre-wrath sign appeared in Joel 2:28-31. Note that the list of events described by Joel occur before God's day of wrath. According to Apostle Peter (Acts 2:1-21), the outpouring of the Holy Ghost, manifested by speaking in tongues, marked the beginning of the sign. Jerusalem was its point of origin and scholars have placed that particular Day of Pentecost in the year AD 33. It will reach its culmination with phenomena driven by the largest sign in the solar system: the Sun. Christians speaking in tongues, under the filling of the Holy Ghost, have been serving as a sign to unbelievers.

> Wherefore tongues are for a sign, not to them that believe, but to them that believe not: but prophesying serveth not for them that believe not, but for them which believe. (I Corinthians 14:22)

Some unbelievers will heed such signs by confessing their sins; repenting; undergoing total submersion for water baptism in the name of the Lord Jesus Christ; and frequently getting filled by the Holy Ghost to lead holy and saved lives. One plight they shall be saved from shall be the Day of Wrath.

Many of those who will not heed that sign will experience the culminating signs. The Sixth Seal sequence of events marks a transition in dispensation. Grace shall diminish and catastrophes from the execution of wrath shall intensify. God compares the Day of Wrath to the sensations of escaping from a lion to be attacked by a bear:

> Woe unto you that desire the day of the Lord! to what end is it for you? the day of the Lord is darkness, and not light. As if a man did flee from a lion, and a bear met him; or went into the house, and leaned his hand on the wall, and a serpent bit him. (Amos 5:18-20)

Why is an attack by a bear worse than an attack by a lion? Lions normally kill their prey as quickly as possible by collapsing their throat.

Bears mall and torture their food. That is the terrifying analogy God selected to depict His Day of Wrath. You do not want to be there!!

Why does God cast creation into His Day of Wrath? The answer was given to the Prophet Isaiah:

> [9.] Behold, the day of the Lord cometh, cruel both with wrath and fierce anger, to lay the land desolate: and he shall destroy the sinners thereof out of it. [10.] For the stars of heaven and the constellations thereof shall not give their light: the sun shall be darkened in his going forth, and the moon shall not cause her light to shine. [11.] And I will punish the world for their evil, and the wicked for their iniquity; and I will cause the arrogancy of the proud to cease, and will lay low the haughtiness of the terrible. [12.] I will make a man more precious than fine gold; even a man than the golden wedge of Ophir. [13.] Therefore I will shake the heavens, and the earth shall remove out of her place, in the wrath of the Lord of hosts, and in the day of his fierce anger. (Isaiah 13:9-13)

Some believe the description of the Day of Wrath in Isaiah has been fulfilled by historic events. Such assertions ignore the fact that history has not recorded an event where the Sun, Moon, and stars had ceased to give their light. Obscuration by clouds will not be the cause of those celestial objects to darken. The Sixth Seal events may be interpreted metaphorically, but they shall be physically experienced.

Joel described it first. Peter recited it to describe the physically observed phenomena of saints speaking in tongues due to the baptism of the Holy Ghost. During the last centuries, events have transpired that made the physical interpretation of Joel 2:30 feasible. Mankind has witnessed wonders in heaven such as the Carrington Event of 1859; the Leonid meteor shower of 1883; and relativistic jets. The subduction regions of tectonic plates; the Grand Canyon; and subterranean aqueous fluids have been among the wonders in the earth. Many medical advances in treating diseases stemmed from man's growing understanding of blood chemistry. Pillars of smoke, not clouds of smoke, have appeared as early 1242 in the form of columns of smoke from rocket exhaust.

The more awe inspiring pillars of smoke appeared in the last half of the twentieth century as the stem of mushroom clouds formed after nuclear and thermonuclear explosions. With the bulk of Joel's prophecy evinced by physical phenomena, there are no solid reasons for assuming the remainder of it would be fulfilled symbolically.

With respect to the rapture, I Thessalonians 5:9-10 indicated the saints have not been appointed to wrath. Since the opening of the Sixth Seal transpires before the Day of Wrath, the rapture may be delayed to happen at some time along the fulfillment of Revelation 6:12-15. Popular depictions of the rapture are disruptive: airliners falling; profusion of traffic accidents; and trains derailing because of the sudden transformations and teleportations to the Lord. If the rapture occurred after the world populations had made it to their caves, cliffs, and dens, the saints' departure would not have caused death and destruction. Just imagine, those left behind would see people instantly glow with intense light and disappear within a twinkling of the eye. How do you think they would feel?

The Fifth Seal events (Revelation 6:9-11) depicted saints, who had been killed for their testimony of God, asking Him how long would He would wait to avenge their blood on those who dwell on earth. The Sixth Seal events then follow. In the first eight verses of chapter seven of the Book of Revelation, the 144,000 servants of the Lord receive their seals. The remainder of that chapter described a large multitude standing before the throne of God with palms in their hands. John is told the great multitude in white robes had come out of great tribulation. The abrupt appearance of multitudes around the throne of God may have been due to the rapture. The bombardment of the Earth from Heaven commences with the opening of the Seventh Seal in chapter eight.

Knowledge of the Lord will not save several people who will be left behind by the rapture. According Revelation 9:20, many of those who shall survive the future judgments will worship devils and idols made of the gold, silver, brass, stone, and wood. Those individuals knew they had survived the Sixth Seal events to inter the Day of Wrath. Yet they will decide not to worship the true God.

Bottom line, everyone will correctly interpret the signs – the Day of Wrath of the Lamb is upon them.

Chapter 9
The Moral

> For God hath not appointed us to wrath, but to obtain salvation by our Lord Jesus Christ, Who died for us, that, whether we wake or sleep, we should live together with him. (I Thessalonians 5:9-10)

This book has provided evidence for treating the physical interpretation of Revelation 6:12-17 as plausible, not preposterous. The New Testament, especially writings by Luke, has been deemed to be historically trustworthy. That fact justifies biblical accounts of the darkness at the crucifixion to be records of an acute solar darkening event (ASDE). The literature survey summarized in Section 2-5 provided evidence for twelve additional ASDE's. Additional evidence of their realty was in the form impulsive nitrate concentration peaks that fell close to the dates of those ASDE's. Their existence suggests severe radiation storms had caused them. Sunspots tell us a blacken Sun will not be a quiet Sun. Subsequently, one should expect the solar blackout to intensify geomagnetic and cosmic ray storms. The Moon glowing blood red could be explained in terms of thermoluminescence caused by the solar cosmic rays bombarding its surface. The intensified radiation could cause the (a) night sky to appear to roll back like a scroll; (b) incapacitation and/or death of aircrews would yield the appearance of falling lights like stars, of aircraft; and (c) entire world populations to flee to underground shelters and cliffs to avoid the heliophysically induced heart failure. The studies discussed in Section 7-5 indicated the rates of morbidity and mortality caused by sudden cardiac death, acute myocardial infarction, and cerebral vascular accidents, would increase with intensified cosmic ray activity. Therefore, cosmic ray induced heart and nervous system diseases, not acute radiation sickness, would be the likely terrifying and kill mechanisms. Intensified geomagnetic activity would provoke negative emotional background phenomena and would induce electrical currents within subterranean aqueous fluids that would be strong enough to dislocate mountains. The seismic phenomena would not destroy the cliffs and underground shelters. Conditions before the opening of the Sixth Seal and breakthroughs in tunnel boring technology will provide the volume of accommodations large enough to shelter the entire world population. And, everyone will attribute this series of events as a sign heralding the Day of Wrath of the Lamb.

Ramifications of the solar blackout could extend beyond the events of the pre-wrath sign. On July 16, 1994, the first in the series of twenty-one large comet fragments hit the planet Jupiter. The energy released by the collision of just one fragment was equivalent to the explosion of 6 million megatons of TNT. The six days of bombardment evinced the risk of similar impacts for the Earth. Subsequently, programs were created to detect and characterize objects that orbit the Sun along paths that are close to the Earth. Other programs were created to examine various mechanisms for destroying and/or diverting near Earth objects (NEO). The NEO detection programs rely on sunlight. Images of NEO's would approach invisibility during the solar blackout. The intensified fields carried in the CMEs and/or high speed solar wind streams of the blacken Sun could magnetically perturb the iron and nickel asteroids into new orbits that would result in collisions with the Earth. The closet groups of near Earth asteroids that could be perturbed are the Atira, Atens, Apollos, and Amors. If the blackout continues for a considerable length of time, an asteroid could approach the Earth undetected to fulfill Revelation 8:8.

The relationship between the powerful electromagnetic surges from the Sun's transition and earthquakes provides an explanation for other scriptures. The long series of concussive ground movements from the mountain as God annunciated the Ten Commandments follows a similar pattern. Scorched rocks on top of Jebel al-Lawz may evince the presence of high voltage plasma. The telluric currents it induced could have caused the subterranean waters to expand and shake the mountain. Most earthquakes transpire within seconds.

The greatest attribute of the heliophysical explanations of the Sixth Seal events that has been presented in this book is its ability to elaborate on the phenomena that forces the world populations to evacuate the surface to underground shelters. Tunnel boring technology has promoted the growth of subterranean facilities. This, with the other descriptions, captures the imagination and causes one to think about the pre-wrath sign. Paul's warning captures the significance of interpreting Revelation 6:12-17 physically:

> For yourselves know perfectly that the day of the Lord so cometh as a thief in the night. For when they shall say, Peace and safety; then sudden destruction cometh upon them, as travail upon a woman with child; and they shall not escape. But ye, brethren, are not in darkness, that that day should overtake you as a thief. Ye are all the children of light, and the children of the day: we are not of the night, nor of darkness. Therefore let us not sleep, as do others; but let us watch and be sober. For they that sleep sleep in the night; and they that be drunken are drunken in the night. But let us, who are of the day, be sober, putting on the breastplate of faith and love; and for an helmet, the hope of salvation. For God hath not appointed us to wrath, but to obtain salvation by our Lord Jesus Christ, Who died for us, that, whether we wake or sleep, we should live together with him. (1 Thessalonians 5:2-10)

This report on the plausibility of physically interpreting Revelation 6:12-17 closes with the results of the crowd witnessing the behavior of the 120 believers who had just been filled with the Holy Ghost:

> Then Peter said unto them, Repent, and be baptized every one of you in the name of Jesus Christ for the remission of sins, and ye shall receive the gift of the Holy Ghost. For the promise is unto you, and to your children, and to all that are afar off, even as many as the Lord our God shall call. And with many other words did he testify and exhort, saying, Save yourselves from this

untoward generation. Then they that gladly received his word were baptized: and the same day there were added unto them about three thousand souls. (Acts 2:38-41)

Index

Symbols

9th hour 28, 29, 30

A

achondrites 44
Acts of John, The 10
acute myocardial infarction 108, 111, 123, 156, 157
acute radiation syndrome 104
aerodynes 59, 60
Africanus, Julius 15
aircraft 58, 59, 60, 61, 63, 64, 103, 104, 105, 106, 123, 146
aircrews 58, 63, 66, 103, 104, 123, 154, 155
airplanes 58, 105, 146
Alaska 3
amateur astronomers xiii, 52
Anchorage, Alaska 3
Anglo-Saxon Chronicle 21, 28, 138
Apocryphal Gospel of Peter 12
Apollo Program 38, 76
apparent magnitude 19
aqueous fluids 85, 86, 89, 90, 91, 120, 124
Aristides xv, 25, 26, 27
Asia Minor 26, 34
asteroid 124
astor 58
Augsburg, Germany 28, 29, 34
aurora borealis 39

B

Baghdad 29, 34
Battle of Plataia 26
Battle of Salamis 26, 27
Battle of Thermopylae xv, 26
Bible verses
   1 Thessalonians 5:2 2
   1 Thessalonians 5:2-10 125
   1 Thessalonians 5:9-10 121, 123
   2 Peter 3:5 37, 89
   2 Timothy 3:16 xvi, 12
   Acts 2:4 17
   Acts 2:12-20 17
   Acts 2:20 21, 33
   Acts 2:38-41 126
   Acts 10:9 10
   Acts 17:34 15
   Acts 19:35 33
   Amos 5:18-20 119
   Amos 8:9 12, 16, 17
   Ecclesiastes 1:9 xv, 21, 23, 84
   Exodus 12:1-14 14
   Exodus 20:4 86
   Genesis 1:2 37
   Genesis 1:9-10 89
   Genesis 7:11-12 85
   Isaiah 13:9-10 37
   Isaiah 13:9-13 120
   Job 5:13-14 10
   Joel 2:31 21, 33, 37
   Jonah 2:6 86, 87
   Leviticus 17:11 108

Luke 21:25-26 106, 107
Luke 22:32 13
Luke 23:44-45 10, 13
Mark 13:24 21, 37
Mark 15:33 10, 12
Matthew 8:24 2
Matthew 24:4-8 5, 98
Matthew 24:29 21
Matthew 27:45 10, 12
Matthew 28:2 14
Revelation 6:12 2, 5, 8, 21, 31, 33, 35, 37, 38, 57, 66, 75, 105
Revelation 6:12-17 xi, xii, xiii, xv, 19, 33, 49, 77, 123, 125
Revelation 6:13 xv, 55, 56, 57, 58, 59, 60, 62, 63, 105
Revelation 6:14 67, 81, 82
Revelation 6:15 xiv, 36, 93, 99, 107, 112
Revelation 6:16-17 117
Revelation 7:1, 3; 8:7; 9:4 7
Revelation 8:8 97, 124
Revelation 8:12 23
Revelation 16:18-20 5
Revelation 16:20-21 84
birthpangs 5
Bithynia 11, 34
black as sackcloth 1, 7, 35, 37, 66
blast 57, 88, 96, 97, 152
blood 1, 7, 14, 17, 18, 37, 38, 53, 58, 103, 106, 108, 109, 110, 118, 120, 121, 123, 147
bondman 93
Book of Apology Against the Heathen 11
Brahe, Tycho 27
Brown, Walt 84

C

Cairo, Egypt 34
California 3, 75, 95
Cameron, Winifred Sawtell 42, 46, 50, 52, 142, 143, 144
cancer 104, 108, 154, 155
Carrington Event 69, 75, 77, 106, 120
Carrington, Richard 70, 71, 74, 148

Catholic Church 14, 135
cave 94
Cesena, Italy 30, 34
Ch'ang-an, China 27, 28, 34
Chiang-hsi province, China 31, 34
chief captains 93
Chile 4, 5, 50
chlorophyll 7, 8
Ch'ü-fu, China 24, 25, 34
Ch'un-ch'iu Chronicle 21, 24, 25
cliffs xii, xiv, xv, 95, 96, 107, 118, 121, 123, 124
CME (coronal mass ejection) 53, 75, 76, 83, 103
Codex Alexandrinus 13
Codex Sinaiticus 13
Codex Vaticanus 13
Coimbra, Portugal 29, 34
comet 39, 124
Constantinople 30, 34
crochet 70
crucifixion xi, xv, 8, 9, 10, 11, 13, 14, 15, 16, 17, 18, 19, 21, 22, 23, 24, 27, 31, 32, 33, 75, 118, 123, 134, 135, 136, 137
Cyril of Jerusalem 14

D

Danjon, Andre-Louis 39, 40
Day of Pentecost 17, 37, 119
Day of the Lord xii, 2
Day of Wrath xii, xv, 37, 97, 112, 117, 119, 120, 121, 124
debris xii, xv, 6, 7, 56, 57, 59, 69, 81, 85, 97, 107
Deccan Flood Basalts 90, 151
den 94
Dionysius the Areopagite 15

E

Earth xii, xiv, 1, 2, 3, 4, 12, 15, 21, 22, 23, 29, 32, 33, 37, 38, 40, 48, 50, 53, 55, 56, 57, 64, 65, 67, 68, 70, 75, 76, 82, 83, 85, 88, 89, 90, 95,

98, 102, 104, 106, 107, 117, 121, 124, 133, 134, 137, 138, 139, 144, 146, 148, 150, 151, 152, 154, 155, 156
earthlights 57, 58
earthquake xii, xv, 1, 2, 3, 4, 5, 6, 7, 8, 10, 11, 13, 14, 28, 37, 57, 58, 81, 82, 88, 113, 133, 135, 138, 149, 150
eclipse, lunar 17, 27, 38, 39, 40, 48, 50, 144
eclipse, solar xi, xiv, 10, 13, 15, 16, 17, 21, 24, 25, 27, 28, 29, 30, 31, 32, 75, 136, 137
  maximum duration 15
  path of totality 29, 31, 75
Egypt 14, 30, 34
electromagnetic radiation 58
EMP (electromagnetic pulse) 63, 64, 65, 147
enstatite 44, 45, 47, 49, 143
Eponym Canon Eclipse 16, 17
eruption 3, 4, 6, 23, 24, 51, 141, 150
eschatology 4, 18
Eusebius 10, 12, 25
explosion 6, 64, 98, 124

F

falling stars 55, 59, 60, 61, 145
flares 43, 44, 47, 48, 49, 53, 65, 68, 70, 74, 76, 141, 143, 148, 149
Florence, Italy 29, 34
Fraunhofer 41, 42
Fu-chien province, China 31, 34

G

Galilei, Galileo 35
geomagnetic xv, 22, 33, 44, 52, 75, 76, 77, 83, 91, 105, 106, 107, 108, 111, 112, 123, 124, 138, 144, 149, 155, 156, 157
GIC (geomagnetically induced currents) 64
Global Position System 82

God xi, xiv, xvi, 1, 12, 14, 15, 16, 18, 37, 84, 86, 87, 88, 89, 119, 120, 121, 123, 124, 125, 133
Gospel of Nicodemus 10, 13
Gospel of Peter 10, 12
Greece 15, 25, 26, 27, 32, 108, 110
Greenacre, James A. xiii, 45, 46, 48, 50, 143
Greenland 5
Ground Level Enhancement (GLE) 102, 103, 110, 154

H

Hawaii 4, 82
heaven xv, 1, 2, 18, 55, 56, 57, 58, 67, 68, 69, 81, 84, 85, 86, 88, 106, 120, 121
Heilsbronn, Germany 28, 34
Hekla 23, 24
heliophysical 32, 125, 156
Herodotus xv, 24, 25, 26, 27, 41, 48, 51, 139
Herschel, Sir William xiii, 39, 40, 46, 141
Hessdalen Valley 57
Historical Eclipses and Earth's Rotation 21, 29, 33, 137, 139
Holy Bible xi, xii, 2, 8, 9, 13, 14, 16, 37, 38, 86, 94
Holy Ghost xiv, 17, 37, 119, 120, 125
hours 12, 14, 29, 30, 31, 41, 47, 48, 51, 53, 57, 61, 69, 70, 74, 85, 94, 105
Humphreys, D. Russell 17, 89, 137, 151
Hydroplate Theory 84, 85, 89, 90, 91
Hα (Hydrogen alpha) 35, 42, 45

I

ICBM 56
impulsive nitrate concentration 77, 123
INDEPTH 89, 151
ionosphere 65, 70
IRBM 57

J

Japan Meteorological Agency 133
Jerusalem xii, 11, 14, 16, 18, 34, 55, 93, 119, 136
Jesus Christ xiv, 1, 2, 8, 14, 23, 27, 31, 38, 75, 87, 98, 101, 117, 119, 123, 125
JLN Laboratory 60
John, Apostle 97
Jupiter 11, 73, 124

K

Kerkrade, Netherlands 28, 29, 34
Kidger, Mark 13, 135, 136
Kopal, Zdeněk xiii, 45, 46, 47, 48, 49, 50, 51, 141, 142, 143
Kozyrev, Nikolai 41, 42, 43, 50, 142
Krakatau 6

L

Link, Frantisek 40
Lisbon, Portugal 3, 4
Louisville, Kentucky 3, 133
Lucca, Italy 29, 30, 34, 107, 140
Lucian 12
lunar rocks 51, 52
lunar transient phenomena 51, 52, 144

M

magnetotelluric 89, 90, 151
Marola, Italy 30, 34
Matsushima, Satoshi 49, 50, 143, 144
Medieval Maximum 28, 31, 34
Mediterranean Sea 87, 90
meteor xii, xv, 7, 33, 46, 55, 56, 67, 81, 102, 120
meteor shower 33, 55, 56, 120
method of line depths 41, 42, 43
Middlehurst, Barbara M. 50, 141, 142, 144
migration 94
missiles 1, 56, 153
Mississippi River 3

Montpelliar, France 29, 34
Moon as blood 37
Moses 14, 37, 88
mountain 5, 6, 67, 81, 82, 84, 87, 88, 89, 90, 96, 124

N

neutron xv, 65, 98, 103, 105, 107, 108, 109, 110, 111, 154, 156, 157
New Madrid, Missouri 3
New Testament 8, 9, 33, 123, 134, 135
Niagara Falls 56
Nicaea, Bithynia 34
Nicodemus 10, 13
Noah 15, 84, 85, 86, 90, 91
noon 4, 10, 12, 16, 17, 29, 30, 140
nuclear war 7, 56, 57
nuclides 52

O

obscuration 17, 23, 30
observatory 23, 41, 42, 43, 45, 48, 49, 50, 51, 69, 71, 72, 83, 100, 135, 144
Olivet discourse 117
Olympiad 10, 11, 16, 25
Oregon 3, 95, 148, 152
Orosius 10, 13, 14, 15, 134, 135, 136

P

Padua, Italy 39, 42
Passover 11, 14, 16
penumbra 17, 35, 38
Perry-Hsu solar activity model 24, 25, 27, 28, 29, 30, 31, 32
Philippines 4
Phlegon 10, 11, 12, 13, 14, 16
phosphorescence 41
phosphors 40
photoelectric 43, 49, 142
photometric 42, 49, 50, 51, 142, 144
photosphere 35
photosynthesis 7, 8, 56
Pien, China 34

Planetarium, Rakurakuen 48
Pliny 9, 25
Plutarch xv, 9, 25, 26, 27
polarimetric 51
Pye Wacket Project 59

R

radial point 56
rapture 117, 121
redden 17, 46, 57
Reichersberg, Austria 28, 29, 30, 34
Repulsin 62
Roman Maximum 15, 27, 34

S

sackcloth 1, 7, 35, 37, 53, 65, 66
Salzburg, Austria 28, 29, 34
Schauberger, Viktor 61, 62, 63, 147
seismic 4, 5, 6, 14, 22, 32, 57, 75, 83, 88, 89, 90, 124, 133, 138, 149, 151
Sekiguchi, Naosuke 51, 52, 144
Seven Books of History Against the Pagans 10, 134
Seventh Day Adventists 4
Shanxi Province, China 56
shelter 93, 96, 98, 107, 124
Siena, Italy 29, 34
Sixth Seal xi, 1, 2, 3, 4, 5, 6, 20, 35, 56, 58, 66, 67, 77, 78, 81, 82, 91, 93, 94, 95, 96, 97, 98, 101, 103, 104, 106, 107, 108, 112, 113, 114, 117, 118, 119, 120, 121, 124, 125
Skylab 67
solar
    blackout xii, xv, 7, 8, 15, 37, 53, 67, 83, 123, 124
    flare 43, 51, 65, 68, 70, 74, 77, 143, 148
    wind 44, 50, 52, 53, 102, 124
solar output model 22, 138
solar proton event 76, 77, 102, 150
Songdo, Korea 31, 34
Son of man 87
South America 4

Split, Croatia 30, 34, 107
Stade, Germany 30, 34
Starfish test 64
stars xv, 7, 11, 12, 13, 19, 23, 29, 30, 32, 33, 51, 55, 56, 57, 58, 59, 60, 61, 106, 120, 123, 137, 141, 144, 145
starspot xii
Stephenson, F. Richard 17, 21, 24, 27, 28, 29, 30, 31, 33, 83, 137, 138, 139, 140, 141, 150
Stone of Calvary 14
Stoupel, Elyiahu 107, 109, 110, 111, 155, 156, 157
sudden cardiac death xv, 108, 110, 123
sun xii, xiii, xv, 1, 2, 6, 7, 8, 11, 12, 13, 15, 16, 17, 18, 19, 20, 21, 22, 23, 24, 25, 26, 27, 28, 29, 30, 31, 32, 33, 35, 36, 37, 38, 40, 43, 53, 56, 57, 58, 64, 65, 66, 68, 69, 71, 72, 73, 74, 75, 82, 83, 89, 91, 102, 104, 106, 107, 108, 110, 112, 118, 119, 120, 123, 124, 134, 136, 139, 140, 141, 148, 149, 150, 155, 156, 157
Sung-chiang, China 30, 31, 34
sunspots xii, 19, 20, 35, 39, 40, 68, 71, 75, 82, 123, 138
supercritical water 85

T

Tambora 6
tau neutrino 32
Tesla, Nikola 63, 147
Tesla pump 62, 63
Thales 24, 25
Thallus 12, 15
Thebes 34
Tiantai, China 30, 31, 34
Toledo, Spain 29, 34
Tractatus de spera 15
transmutation 38, 89
trigger 5, 22
tsunami 4, 149
turbine 61, 63, 147

131

U

umbra 35, 38, 73
underground facilities xv, 82, 94, 96, 97, 99, 100, 101, 153
universes, parallel 133
University of Manchester 42, 44, 46
U.S. Geological Survey 152

V

Vanguard 83
Venus 11, 29
Vigeois, France 29, 34

W

warheads 1, 7, 56, 57, 96
water 2, 3, 37, 38, 62, 63, 70, 83, 84, 85, 86, 87, 88, 89, 90, 95, 119, 150, 151
White, Ellen G. 4, 55, 133, 145
WR104 66, 98
Wrath of the Lamb xii, 112, 117, 121, 124

Y

Yohkoh 8

Z

Zwiefalten, Germany 28, 29, 34

Endnotes

[1] Tegmark, M. (2004). Parallel universes. In, J. D. Barrow, P. C. W. Davies, & C. L. Harper (Eds.), *Science and Ultimate Reality: Quantum Theory, Cosmology, and Complexity* [pp. 459-491]. London: Cambridge University Press. Available as preprint arXiv: astro-ph/0302131.

[2] Nyffenegger, P. (1997, April 18). Earthquake pages. Retrieved June 19, 2002, from http://wadati.ig.utexas.edu:8000/earthquake/EQFAQ/worldbody.html.

[3] White, E. G. (1988). *America in prophecy*. Jemison, AL: Inspiration Books East, Inc. (Original work published 1888 under the title The Great Controversy). Signs of the end retrieved February 15, 2002 from the The Ellen G. White Web Site at http://www.ellenwhite.org/egw42.htm.

[4] Ludman, A., & Coch, N. K. (1982). *Physical geology* (p. 435). New York: McGraw-Hill Book Company

[5] Bullen, K. E. (1970). Earthquake. In *Encyclopedia international* (Vol. 6, pp. 186-189). New York: Grolier, Inc.

[6] Japan Meteorological Agency (2000). Recent seismic activity in the Miyakejima and Niijima-Kozushima region, Japan – The largest earthquake swarm ever recorded. *Earth, Planets and Space*, **52**(8), i-viii. Terra Scientific Publishing Company, Tokyo.

[7] Altschuler, M. D., (1969). Chapter 7. Atmospheric electricity and plasma interpretation of UFOs. In E. U. Condon (Sci. Dir.) & D. S. Gillmor (Ed.). *Scientific study of Unidentified Flying Objects* (pp. 723-755). New York: A Bantam Books, Inc

[8] Brown, W. (2001). *In the beginning: Compelling evidence for creation and the flood* (7th ed., pp. 105,106, 116, 138-149). Phoenix, AZ: Center for Scientific Creation.

[9] Taylor, C. E. (1979). *World War III and the destiny of America* (p. 325). Nashville, TN: Thomas Nelson Inc., Publishers.

[10] Jensen, W. A., Heinrich, B., Wake, D. B., Wake, M., & Wolfe, S. L. (1979). *Biology* [p. 65]. Belmont, CA: Wadsworth Publishing Company.

[11] Polkinghorne, J., & Beale, N. (2009). *Questions of Truth: Fifty-one Responses to Questions about God, Science, and Belief* [pp. 85-88]. Louisville, KY: Westminster John Knox Press..

[12] Blomberg, C. L. (2007). *The Historical Reliability of the Gospels* [p. 304]. Downers Grove, IL: InterVarsity Press.

[13] Blomberg, p. 11.
[14] McDowell, J. (1999). *The New Evidence That Demands A Verdict* [pp. xiii-xiv]. Nashville, TN: Thomas Nelson Publishers.
[15] Ibid., pp. 63-66.
[16] Bruce, F. F. (1981). *The New Testament Document: Are They Reliable?* (6th Ed.) [pp. 80-93]. Grand Rapids, MI: William. B. Eerdmans Publishing Company.
[17] Anderson, D. (2007, April 6). Darkness at the crucifixion: Metaphor or real history? Retrieved March 3, 2009 from http://creation.com/darkness-at-the-crucifixion-metaphor-or-real-history.
[18] Blomberg, p. 251.
[19] Weston, W. (1748). On the miraculous darkness at our saviour's Passion. *Dissertations on Some of the Most Remarkable Wonders of Antiquity* [pp. 1-62]. London, Cambridge: F. Bentham, Printer to the University. Retrieved July 31, 2009 from http://books.google.com.
[20] Hewer, D. (2009). *The Historical Reliability of the New Testament* (2nd Ed.) [p. 46]. Retrieved August 10, 2009, from www.WhyFaith.com/nt.
[21] Blomberg, pp. 251-256.
[22] Brewer, B. (1991). *Eclipse* (2nd ed.) [pp. 14]. Seattle, WA: Earth View, Incorporated.
[23] Habermas, G. (2005). Recent perspectives on the reliability of the Gospels. *Faculty Publications and Presentations*. Liberty Baptist Theological Seminary and Graduate School. Liberty University. Retrieved October 30, 2009, from http://digitalcommons.liberty.edu/its_fac_pubs/106.
[24] Ehrman, B. D. (Ed.) (2003). *Lost Scriptures – Books that Did Not Make It into the New Testament* [pp. 32-33].
[25] Barnstone, W. (Ed) (1984). *The Other Bible* [p. 368]. New York, NY: HarperCollins.
[26] Barnstone, p. 419.
[27] Fotheringham, J. K. (1920, December). A solution of ancient eclipses of the sun. *Monthly Notices of the Royal Astronomical Society*, **81**(2), 104-126.
[28] Orosius, P. (A.D. 417). *The Seven Books of History Against the Pagans*. In, R. J. Deferrari (Trans.) & H. Dressler, et al. (Vol. Eds.) (1964). *The Fathers of the Church – Vol. 50* (1st short-run printing 2001, pp. 291-292). Washington, DC: The Catholic University of America Press.
[29] Barnett, J. E. (1998). *Time's Pendulum – The Quest to Capture Time – From Sundials to Atomic Clocks* [p. 45]. New York, NY: Plenum Press is a Division of Plenum Publishing Corporation.
[30] Savile, B. W. (1858). *The First and Second Advent: Or the Past and the Future* [pp. 30-33]. London: Wertheim, Macintosh, & Hunt. Retrieved August 4, 2009 from http://books.google.com.
[31] Stockwell, J. N. (1892, November). Supplement to recent contributions to chronology and eclipses. *Astronomical Journal*, **12**(no. 16, issue 280), 121-125.
[32] Weston, p. 4.
[33] Finegan, J. (1999, October). *Handbook of Biblical Chronology* (Revised Edition – Second Printing) [p. 97]. Peabody, MA: Hendrickson Publishing.

34. Fotheringham, pp. 104-126.
35. Finegan, pp. 360-369.
36. Missler, C. (1999). *Cosmic Codes* [pp. 233-237]. Cour d'Alene, ID: Koinonia House.
37. McDowell, pp. 197-201.
38. Stockwell.
39. Courtenay, R (1911, June). Moon's visibility and the date of the crucifixion. *The Observatory*, **34**(436), 228-232.
40. Pratt, J. P. (1991, September). Newton's date for the crucifixion. *Royal Astronomical Society Quarterly Journal*, **32**(3), 301-304.
41. Ibid.
42. Dodgson Dodgson, C. (Trans.)(1842). Tertullian. Vol. I. Apologetic and Practical Treatises [pp. 50-51]. In, *A Library of Fathers of the Holy Catholic Church*. Oxford, John Henry Parker: J. G,. F. and J. Rivington, London. Retrieved July 16, 2009, from http://books.google.com.
43. Wace, H., & Piercy, W. C. (Eds.) (1999). *Dictionary of Early Christian Biography* [p. 943]. Hendrickson Publishers, Inc.
44. Chambers, G. F. (1904). *The Story of Eclipses* [p. 110]. New York: D. McClure, Phillips & Company. First printing 1899. Retrieved May 29, 2009, from http://books.google.com.
45. Ussher, J., & Pierce, L. (Trans.)(2007). *Annals of the World* [p. 822]. Green Forest, AR: New Leaf Publishing Group
46. Weston, p. 22.
47. Bruce, pp. 80-93.
48. Aveni, A. F. (1995). *Empire of Time: Calendars, Clocks, and Cultures* [pp. 90-92]. New York: Kodansha America, Incorporated.
49. Duncan, D. E. (1998). *Calendar: Humanity's Epic Struggle to Determine a True and Accurate Year* [pp. 47-48]. New York, NY: Avon Books, Incorporated.
50. Blomberg, p. 28.
51. Connolly, R. H (Trans.) (1929). *Didascalia Apostolorum*. Oxford: Clarendon, Press. Retrieved October 6, 2009 from http://www.bombaxo.com/didascalia.html. Extent of darkness at the crucifixion and the earthquake at the resurrection are mentioned in Chapter XXI, verse 15.
52. Ehrman, B. D. (Ed.) (2003). *Lost Scriptures – Books that Did Not Make It into the New Testament* [pp. 32-33]. New York, NY: Oxford University Press, Incorporated.
53. Kidger, M. (1999). *The Star of Bethlehem: An Astronomer's View* [pp. 68-72]. Princeton, NJ: Princeton University Press.
54. Finegan, pp. 363-364.
55. Rohrbacher, D. (2002). *The Historians of Late Antiquity* [p. 138-149]. New York, NY: Rutledge - Taylor & Francis Group.
56. Wace, Ref. 23, pp. 793-794.
57. Orosius, pp. 291-292].
58. *Concordant Version - The Sacred Scriptures – New Testament: Greek Text with*

English Sublinear and Supperlinear (1955). Saugus, CA: Concordant Publishing Concern. (Pages 27, 118, 177, 274, 700).

[59] Pallotta, C. (1995). *The Crucifixion Eclipse* (pages 2, 4). Brooklyn, NY: Marian Media Apostolate. He cites Merk, A. S. J. (Ed.) (1951). *Novum Testamentum Graece Et Latine*, (pages 104, 181, 298, 307). Romae: Sumptibus Pontificii Instituti Biblici.

[60] Thiede, C. P., & d'Ancona, M. (1996). *The Jesus Papyrus* [p. 136]. New York, NY: A Galilee Book published by Doubleday, a division of Random House, Incorporated.

[61] Orosius, pp. 291-292.

[62] Ehrman, pp. 32-33.

[63] Pick, B. (1887). *The Life of Jesus According to Extra-Canonical Sources* [pp. 142-143]. New York, NY: John B. Alden, Publisher. Retrieved July 29, 2009, from http://books.google.com.

[64] Dodgson, p. 51.

[65] Ambraseys, N. (2005, July). Historical earthquakes in Jerusalem – A methodological discussion. *Journal of Seismology*, **9**(3), 329-340.

[66] Ibid.

[67] Adriatikus (2008, May 1). Communiqué and photograph to the author.

[68] Thiede, C. P. (2005). *The Emmaus Mystery* [p. 178]. New York, NY: Continuum International Publishing.

[69] Trench, G. H. (1908). *The Crucifixion and Resurrection of Christ by the Light of Tradition* [p. 113]. London, England: John Nurray, Publisher.

[70] Habermas, G. R. (1996). Ancient non-Christian sources. *Faculty Publications and Presentations*. Liberty Baptist Theological Seminary and Graduate School. Retrieved July 21, 2009, from http://digitalcommons.liberty.edu/lts_fac_pubs/39

[71] Missler, pp. 251-252.

[72] Eddy, J. (1977, June). Climate and the changing sun. *Climatic Change*, **1**(2), 173-190..

[73] Bartlett, G. F. (2008). *The Natural and Supernatural in the Middle Ages* [pp. 51-53]. New York: Cambridge University Press.

[74] Bartlett, pp. 67-69.

[75] Wace, H., & Piercy, W. C. (Eds.) (1999). *Dictionary of Early Christian Biography* [pp. 260-261]. Hendrickson Publishers, Inc.

[76] Meeus, J. (2003, December). The maximum possible duration of a total solar eclipse. *Journal of the British Astronomical Association*, **113**(6), 343-348.

[77] Bruce, pp. 116-117.

[78] Adler, W., & Tuffin, P. (2002). *The Chronography of George Synkellos: A Byzantine Chronicle of Universal History from the Creation* [p. 466]. Oxford Press. Retrieved March 12, 2008, from http://www.tertullian.org/rpearse/cyncellus/.

[79] Wace, p. 8.

[80] Chambers, p. 110.

[81] Fotheringham, p. 112.

[82] Kidger.

83. Hind, J. R. (1872). Historical eclipses. *Astronomical Register*, **10**, 207-214.
84. Sawyer, J. F. A. (1972). Why is a solar eclipse mentioned in the passion narrative (Luke XXIII. 44-5)? *The Journal of Theological Studies*, **23**(1), 124-128.
85. Brewer, B. (1991). *Eclipse* (2nd ed.) [p. 17]. Seattle, WA: Earth View, Inc.
86. Lynn, W. T. (1909). *Remarkable Eclipses: A Sketch of the Most Interesting Circumstances Connected with the Observation of Solar and Lunar Eclipses, Both in Ancient and Modern Times* (10th Edition, Revised) [pp. 8-9]. London, England: Samuel Bagster & Sons, Ltd.
87. Fotheringham, p. 107.
88. Stephenson, F. R. (1997). *Historical Eclipses and Earth's Rotation* [pp. 125-127]. Cambridge, United Kingdom: Cambridge University Press.
89. Hind, J. R. (1872). Historical eclipses. *Astronomical Register*, **10**, 207-214.
90. Johnson, S. J. (1874). *Eclipses, Past and Future; with General Hints for Observing the Heavens* [p. 9]. Oxford and London: James Parker and Company.
91. Humphreys, C. J., & Waddington, W. G. (1985, March). The date of the Crucifixion. *Journal of the American Scientific Affiliation*, **37**, 2-10. Original article published 1983 in *Nature*, **306**, 743-746.
92. Schaefer, B. E. (1990, March). Lunar visibility and the crucifixion. *Royal Astronomical Society Quarterly Journal*, **31**(1), 53-67.
93. Ruggles, C. (1990, June 21). The Moon and the Crucifixion. *Nature*, **345**(6277), 669-670.
94. Schaefer, B. E. (1991, July). Glare and celestial visibility. *Publications of the Astronomical Society of the Pacific*, **103**, 645-660.
95. Walvoord, J. F. (1999). *Every Prophecy of theBbible* [pp. 557-558]. Colorado Springs, CO: Chariot Victor Publishing.
96. Lockyer, H. (1971, December). *All of the Miracles of the Bible* (Eleventh printing, first edition printed 1961) [p. 260]. Grand Rapids, MI: Zondervan Publishing House.
97. Hoyt, H. A. (1969). *The End Times* [pp. 158-159]. Chicago, IL: The Moody Bible Institute of Chicago.
98. Bloomfield, A. E. (2002). *The Key to Understanding Revelation: An Easily Grasped Structure of a Complex Book* [p. 150]. Bloomington, MN: Bethany House Publishers
99. Bennett, J., Donahue, M., Schneider, N., & Voit, M. (2002). *The Cosmic Perspective* (2nd ed.) [pp. 499-506, 515]. San Francisco, CA: Addison Wesley.
100. Perkins, S. (2004, July 31). Parting shots. *Science News*, **166**, 74-76.
101. Menzel, D. H. (1964). *A Field Guide to the Stars and Planets* [p. 117]. Boston, MA: Houghton Mifflin Company.
102. Williams, D. B. (2004, March 3). Private communiqué with the author.
103. Meeus, p. 343.
104. Morrison, L. (2001, May). History in the service of astronomy. *Journal for the History of Astronomy*, **32**(106), 160-162.
105. Stephenson, F. R., & Yau, K. K. C. (1992, February). Astronomical records in the Ch'un-Ch'iu Chronicle. *Journal for the History of Astronomy*, **23**(1), 31-51.

[106] Ingram, J. H. (Trans.) (1823). *The Anglo-Saxon Chronicle.* Retrieved March 1, 2009 from http://www.gutenberg.org/etext/657.

[107] Clark, D.H., & Stephenson, F. R. (1978, December). An interpretation of the pre-telescopic sunspot records from the Orient. *Quarterly Journal of the Royal Astronomical Society,* 19(4), 387-410.

[108] Willis, D. M., Armstrong, G. M., Ault, C. E., & Stephenson, F. R. (2005). Identification of possible intense historical geomagnetic storms using combined sunspot and auroral observations from East Asia. *Annales Geophysicae,* 23, 945-971.

[109] Willis, D. M., & Stephenson, F. R. (2001). Solar and auroral evidence for an intense recurrent geomagnetic storm during December in AD 1128. *Annales Geophysicae,* 19, 289-302.

[110] Yau, K. K. C., & Stephenson, F. R. (1988). A revised catalogue of Far Eastern observations of sunspots (165 BC to AD 1918). *Quarterly Journal of the Royal Astronomical Society,* 29(2), 175-197.

[111] Mazzarella, A., & Palumbo, A. (1988, July-August). Solar, geomagnetic and seismic activity. *IL Nuovo Cimento,* 11-C(4), 353-364.

[112] Palumbo, A. (1989, November-December). Gravitational and geomagnetic tidal source of earthquake triggering. *IL Nuovo Cimento,* 12-C(6), 685-693.

[113] Shaltout, M. A. M., Tadros, M. T. Y., & Mesiha, S. L. (1999). Power spectra analysis for world-wide and North Africa historical earthquake data in relation to sunspots periodicities. *Renewable Energy,* 17, 499-507.

[114] Shatashvili, L. Kh., Sikharulidze, D. I., & Khazaradze, N. G. (2000, January). Dynamics of changes in the IMF sector structure in the vicinity of the Earth and the problem of earthquakes. *International Journal of Geomagnetism and Aeronomy,* 1(4). Retrieved May 16, 2003 from http://www.izmiran.rssi.ru/magnetism/SSIMF/PAPERS/GAI99329/GAI99329.HTM.

[115] Han, Y., Guo, Z, Wu, J., & Ma L. (2004, March). Possible triggering of solar activity to big earthquakes (Ms$\geq$8) in faults with near west-east strike in China. *Science in China Series G: Physics and Astronomy,* 47(2), 173-181.

[116] Perry, C. A., & Hsu, K. J. (2000, November). Geophysical, archaeological, and historical evidence support a solar output model for climate change. *Proceedings of the National Academy of Science,* 97(23), 12433-12438.

[117] Perry, C. A. (2009, June 10). The solar output model was attached to the private communiqué to the author. It was modified to generate annual, instead of decade, irradiance percentages.

[118] Borucki, W. J., et al. (2009, August 7). Kepler's optical phase curve of the exoplanet HAT-P-7b. *Science,* 325(5941), 709.

[119] Basri, G., Borucki, W. J., & Koch, D. (2005, May). The Kepler Mission: A wide-field transit search for terrestrial planets. *New Astronomy Reviews,* 49(7-9), 478-485.

[120] Press Kit (2009, February). *Kepler: NASA's First Mission Capable of Finding Earth-Size Planets* [p. 8]. Retrieved September 22, 2009 from http://www.nasa.gov/kepler.

[121] Clark & Stephenson, 398.

[122] Scuderi, L. A. (1990). Oriental sunspot observations and volcanism. *Quarterly Journal of the Royal Astronomical Society*, **31**(1), 109-120.

[123] Global Volcanism Program web site. Retrieved March 26, 2010, from http://www.volcano.si.edu/world/volcano.cfm?vnum=1702-07=&volpage=erupt.

[124] Stephenson, F. R. (1997). *Historical Eclipses and Earth's Rotation* [pp. 338-341]. Cambridge, United Kingdom: Cambridge University Press.

[125] Stephenson & Yau, pp. 40-41.

[126] Usoskin, I. G., Solanki, S. K., & Kovaltsov, G. A. (2007, August). Grand minima and maxima activity: New observational constraints. *Astronomy and Astrophysics*, **471**(1), 301-309.

[127] Chambers, p. 94.

[128] Stephenson, p. 342.

[129] Stephenson, p. 342.

[130] Chamber, p. 94.

[131] Stephenson, p. 343.

[132] Fotheringham, J. K. (1920, December). A solution of ancient eclipses of the sun. *Monthly Notices of the Royal Astronomical Society*, **81**(2), 104-126.

[133] Stephenson & Yau, pp. 40-41.

[134] Usoskin (2007), p. 5.

[135] Chambers, pp. 97-98.

[136] Dupuy, R. E., & Dupuy, T. N. (1970). *The Encyclopedia of Military History from 3000 B.C to the Present* [pp. 25-26]. New York, NY: Harper & Row, Publishers.

[137] Lynn, W. T. (1909). *Remarkable Eclipses: A Sketch of the Most Interesting Circumstances Connected with the Observation of Solar and Lunar Eclipses, Both in Ancient and Modern Times* (10th Edition) [pp. 11-13]. London: Samuel Bagster & Sons Limited.

[138] Hind, J. R. (1872). Historical eclipses. *Astronomical Register*, **10**, 207-214.

[139] Herodotus, Macaulay, C. C. (Trans.), & Lateiner, D. (Rev.) (2004). *The Histories* [p. 464]. New York, NY: Barnes & Noble Classics.

[140] Hind, p. 209.

[141] Lynn, p. 13.

[142] Chambers, p. 98.

[143] Stephenson, p. 343.

[144] Johnson, S. J. (1874). *Eclipses, Past and Future; with General Hints for Observing the Heavens* [p. 15]. Oxford and London: James Parker and Company.

[145] Usoskin (2007), p. 5.

[146] Stephenson, p. 236.

[147] Eddy, J. (1977, June). Climate and the changing sun. *Climatic Change*, **1**(2), 173-190.

[148] Ingram.

[149] Chambers, p. 122.

[150] Chambers, p. 34.

[151] Stephenson, pp. 245-246.

[152] Stephenson., pp. 245-246.

[153] Ogurtsov, M. G., Nagovitsyn, Y. A., Kocharov, G. E., & Jungner, H. (2002,

December). Long-period cycles of the sun's activity recorded in direct solar data and proxies. *Solar Physics*, **211**(1), 371-394.
[154] Stephenson, p. 251.
[155] Eddy, J. (1977, June). Climate and the changing sun. *Climatic Change*, 1(2), 173–190.
[156] Stephenson, pp. 423-424.
[157] Stephenson, p. 392.
[158] Stephenson, pp. 392-393.
[159] Stephenson, p. 424.
[160] Stephenson, p. 393.
[161] Stepenson, p. 393.
[162] Chambers, p. 124.
[163] Harper, D. (2001). Noon. Online etymology dictionary. Retrieved March 30, 2009, from http://www.etymonline.com/index.php?term=noon.
[164] Stephenson., p. 450.
[165] Ibid., p. 419.
[166] Ibid., pp. 80-81.
[167] Ibid., p. 399.
[168] Ibid., p. 385.
[169] Ibid., p. 400.
[170] Ibid., p. 399.
[171] Ibid., pp. 397-398.
[172] Ibid., p. 400.
[173] Ibid., p. 398.
[174] Kington-Oliphant, T. L. (1875). *The Duke and the Scholar* [p. 96]. London: MacMillan and Co. Retrieved April 10, 2009 from http://books.google.com/books?id=BigMAAAAIAAJ&pg=PA96&lpg=PA96&dq=Lucca+1239+%22Oliphant%22&source=bl&ots=hsbMt-4sg-&sig=Da9x4_TlSGw_eQw-RpDzG5vyvww&hl=en&ei=6KPfSbqhKKmEngfB9tWlCg&sa=X&oi=book_result&ct=result&resnum=1#PPP1,M1.
[175] Stephenson, pp. 401, 403.
[176] Ibid., p. 401.
[177] Harper, D. (2001). Noon. Online etymology dictionary. Retrieved March 30, 2009, from http://www.etymonline.com/index.php?term=noon.
[178] Stephenson, p. 398.
[179] Stephenson, p. 402.
[180] Stephenson, p. 445.
[181] Stephenson, p. 401.
[182] Johnson, p. 56.
[183] Pang, K., Yau, K. K., & Chou, H. (2002). Astronomical dating and statistical analysis of ancient eclipse data. In, S. M. R. Ansari (Ed.). *History of Oriental Astronomy* [p. 98]. Dordrecht, The Netherlands: Kluwer Academic Publishers
[184] Stephenson, pp. 259-260.
[185] Stephenson, pp. 263-264, 265.
[186] Witmann, A. D., & Xu, Z. T. (1987, July). A catalogue of sunspot observations

from 165 BC to AD 1684. *Astronomy & Astrophysics Supplement Series*, **70**, 83-94.

[187] Ogurtsov, M. G., Nagovitsyn, Y. A., Kocharov, G. E., & Jungner, H. (2002, December). Long-period cycles of the sun's activity recorded in direct solar data and proxies. *Solar Physics*, **211**(1), 371-394

[188] Stephenson, pp. 260-261, 274-275.

[189] Riesselmann, K. (2000, Auagust 4). DONUT finds missing puzzle piece. *Ferminews*, **23**(14), 2-5.

[190] Winklhofer, M. (2005). Biogenic magnetite and magnetic sensitivity in organisms – From magnetic bacteria to pigeons [Invited Lecture]. Joint 15th Riga and 6th PAMIR Conference on Fundamental and Applied MHD. Institute of Physics, University of Latvia, Riga, Latvia. Also, available in *Magnetohydrodynamics Journal*, **41**(4), 295-304.

[191] Walker, M. M., Dennis, T. E., & Kirschvink, J. L. (2002). The magnetic sense and its use in long-distance navigation by animals. *Current Opinion in Neurology*, **12**, 735-744.

[192] Paneth, F. A. (1961). History of meteorites. In *Encyclopaedia Britannica* (Vol. 15, p. 341). Chicago: Encyclopaedia Britannica.

[193] Yeomans, D. K. (1991). *Comets: A Chronological History of Observations, Science, Myth, Folklore* [pp. 188-190]. New York, NY: John Wiley & Sons, Inc.

[194] Solanki, S. K. & Unruh, Y. C. (2003, November 13). Spot sizes on Sun-like stars. Preprint arXiv: astro-ph/0311310

[195] Hughes, D., Paczukski, M., Dendy, R.O., et al (2003). Solar flares as cascades of reconnecting magnetic loops. *Physical Review Letters*, **90**(13), 131101. Preprint arXiv: cond-mat/0210201.

[196] Hughes, D., Paczuski, M., Dendy, R. O., Helander, P., & McClements, K. G. (2002, October 9). Solar flares as cascades of reconnecting magnetic loops. Preprint arXiv: cond-mat/0210201.

[197] Kopal, Z. (1966, December). Lunar flares. *Astronomical Society of the Pacific Leaflets*, **9**(450), 1-8. Provided by the NASA Astrophysics Data System.

[198] Herschel, W. (1956, May). Herschel's 'Lunar volcanos.' *Sky and Telescope*, pp. 302-304. (Reprint of An Account of Three Volcanos in the Moon, William Herschel's report to the Royal Society on April 26, 1787, reprinted from his *Collected Works* (1912)).

[199] Middlehurst, B. M. (1964, August). A lunar eruption in 1783? *Sky & Telescope*, **28**(2), 83-84.

[200] Kopal, Z. (1965, May). The luminescence of the moon. *Scientific American*, **212**(5), 28.

[201] Kopal, Z. & Rackham, T. W. (1963). Excitation of lunar luminescence by solar activity. *Icarus*, **2**, 481-500

[202] Kopal & Rackham, 483.

[203] Kopal & Rackham, 484.

[204] Grainger, J.E., &Ring, J. (1962). Lunar luminescence. In Z. Kopal & Z. K. Mikhailov (Eds.), International Astronomical Union, Symposium 14. *The*

[205] Sheehan, W., & Dobbins, T. (1999, September). The TLP myth: A brief for the prosecution," *Sky & Telescope*, **98**(3), 118-123.

[206] Middlehurst, B. M., Burley, J. M., Moore, P. A., & Welther, B. L. (1968, July). *NASA Technical Report R-277: Chronological Catalog of Reported Lunar Events.* Washington, D.C: NASA.

[207] Kozyrev, N. A. (1962). "Spectrographic Proofs for Existence of Volcanic Processes on the Moon," *The Moon* (editors Z. Kopal and Z.K. Mikhailov), International Astronomical Union, Symposium 14, Academic Press Inc., New York, pp. 263-272. Provided by the NASA Astrophysics Data System.

[208] Doel, R. E. (1996, October). The lunar volcanism controversy. *Sky & Telescope*, **92**(4), 26-30.

[209] Gehrels, T., Coffeen, T., & Owens, D. (1964, December). Wavelength dependence of polarization. III. The lunar surface. *Astronomical Journal*, **69**(10), 826-852. Provided by the NASA Astrophysics Data System.

[210] Kopal & Rackham, 485.

[211] Grainger, J.E., &Ring, J. (1962). Lunar luminescence. In Z. Kopal & Z. K. Mikhailov (Eds.), International Astronomical Union, Symposium 14. *The Moon* (pp. 445-452). New York: Academic Press, Inc. Provided by the NASA Astrophysics Data System.

[212] Kopal & RAckham, 487.

[213] Wildey, R., & Pohn, H. (1964, October). Detailed photoelectric photometry of the Moon. *Astronomical Journal*, **69**(8), 619-634. Provided by the NASA Astrophysics Data System.

[214] Cameron, W. S. (1980, March). New results from old data: Lunar photometric anomalies in Wildey and Pohn's 1962 observations. *Astronomical Journal*, **85**(3), 314-328. Provided by the NASA Astrophysics Data System.

[215] Spinrad, H. (1964). Lunar luminescence in the near ultraviolet. *Icarus*, **3**, 500-501.

[216] Wildey, R. (1964, April). Lunar luminescence. *Publications of the Astronomical Society of the Pacific*, **76**(449), 112-114. Provided by the NASA Astrophysics Data System.

[217] Scarfe, C.D. (1965). Observations of lunar luminescence at visual wavelengths. *Monthly Notices of the Royal Astronomical Society*, **130**, 19-29. Provided by the NASA Astrophysics Data System.

[218] Federer, C. A., (1963, November). Auroras in September. *Sky and Telescope*, **26**(5), 256-257.

[219] Federer, C. A., (1963, December). September aurora sequel. *Sky and Telescope*, **26**(6), 322.

[220] Geake, J. E., Lipson, H., & Lumb, M. D. (1962). Laboratory simulation of lunar luminescence. In Z. Kopal and Z. K. Mikhailov (Eds.) *The Moon* (I.A.U. Symposium, no. 14, pp. 441-444). New York: Academic Press. Provided by the NASA Astrophysics Data System.

[221] Derham, C.J., & Geake, J.E. (1964, January 4). The luminescence of meteorites. *Nature*, **201**, 62-63.
[222] Greenacre, J. A. (1963, December). A recent observation of lunar color phenomena. *Sky & Telescope*, **26**(6), 316-317.
[223] Zahner, D. D. (1963-64, December – January). Air force reports lunar changes. *Review of Popular Astronomy*, **57**(525), 29, 36.
[224] Ley, W. (1965). *Ranger to the Moon* [p. 71]. New York: The New American Library of World Literature, Inc.
[225] Jackson, J. H. (1964, October). The lunascan project presents inconsistent Moon. *Analog Science Fiction/Science Fact*. Retrieved March 2003 from http://www.astrosurf.com/lunascan/analog.htm.
[226] Cameron, W. S. (1978, July). *Lunar Transient Phenomena Catalog* (NSSDC/WDC-A-R&S 78-03). Greenbelt, MD: NASA Goddard Space Flight Center. Event Serial No.778.
[227] Kopal, Z. & Rackham, T. W. (1964, March). Lunar luminescence and solar flares. *Sky & Telescope*, **27**(3), 140-141.
[228] Meaburn, J. (1994, June). Z. Kopal (1914-1993). *Quarterly Journal of the Royal Astronomical Society*, **35**, 229-230. Provided by the NASA Astrophysics Data System.
[229] Kopal, Z. & Rackham, T. (1962). Cine photography of the Moon from Pic-Du. In Z. Kopal & Z. K. Mikhailov (Eds.), *The Moon* (I.A.U. Symposium, no. 14, pp. 343-360). New York: Academic Press.
[230] Kopal, Z. (1959, September). Does the moon posses a magnetic field? *Space Journal*, **2**(1), 3-8.
[231] Kopal, Z. & Rackham, T. W. (1964, March). Lunar luminescence and solar flares. *Sky & Telescope*, **27**(3), 140-141.
[232] Ney, E. P., Woolp N. J, & Collins, R. J. (1966, April 1). Mechanisms for lunar luminescence. *Journal of Geophysical Research*, **71**(7), 1787 – 1793.
[233] Ashbrook, J. (Ed.) (1964, January). *Sky and Telescope*, **27**(1), 3.
[234] Ley, W. (1964). *Missiles, Moonprobes, and Megaparsecs* [p. 91]. New York: The New American Library of World Literature, Inc.
[235] Ley, W. (1965). *Ranger to the moon* (p. 71). New York: The New American Library of World Literature, Inc.
[236] Ashbrook, 3.
[237] Matsushima, S., & Zink, J. R. (1964, September). Three-color photometry of Mare Crisium during the total eclipse of 30 December 1963. *Astronomical Journal*, **69**(7), 481-484. Provided by the NASA Astrophysics Data System
[238] Derham, C.J., Geake, J. E., & Walker, G. (1964, July). Luminescence of enstatite achondrite meteorites. *Nature*, **203**, 134-136.
[239] Chanin, M.-L., Lepine, V., & Blamont, J. E. (1982). Thermoluminescence of the lunar surface. *The Moon and the Planets*, **27**, 143-163. Provided by the NASA Astrophysics Data System.
[240] Morgan, T. H. (September 1983). Lunar luminescence measurements [Abstract]. In *The 1983 NASA/ASEE Summer Faculty Fellowship Research Program Research*

Reports (SEE N86-14078 04-85). Johnson (Lyndon B.) Space Center. Provided by the NASA Astrophysics Data System.

[241] Adair, L. P., & Irvine, W. M. (1973, April). Monochromatic phase curves and albedos for the lunar disk. *Astronomical Journal*, 78(3), 267-277. Provided by the NASA Astrophysics Data System.

[242] Sanduleak, N. & Stock, J. (1965). Indications of luminescence in the December 1964 eclipse. *Astronomical Society of the Pacific*, 77, 237-240. Provided by the NASA Astrophysics Data System.

[243] Matsushima, S. (1966, October). "Variation of lunar eclipse brightness and its association with the geomagnetic planetary index K_p. *Astronomical Journal*, 71(8), 699-705. Provided by the NASA Astrophysics Data System.

[244] Middlehurst, B. M. (1966, December). Transient changes in the Moon. *Observatory*, 86(955), 239-242. Provided by the NASA Astrophysics Data System.

[245] Middlehurst, B. M., Burley, J. M., Moore, P. A., & Welther, B. L. (1968, July). *NASA Technical Report R-277: Chronological Catalog of Reported Lunar Events*. Washington, D.C: NASA.

[246] Cameron (1978), Event Serial No. 1029 & 1031.

[247] Moore, P. (1971). Transient phenomena on the Moon. *Quarterly Journal of the Royal Astronomical Society*, 12, 45-47.

[248] Sekiguchi, N. (1971, May). An anomalous brightening of the lunar surface observed on March 26, 1970. *Earth, Moon, and Planets*, 2, 423-434. Provided in the NASA Astrophysics Data System.

[249] Iriarte, B., Johnson, H. L., Mitchell, R. I., & Wisniewski, W. K. (1965, July). Five-color photometry of bright stars and The Arizona-Tonantzintla catalogue. *Sky and Telescope*, 30(1), 21-31.

[250] Sekiguchi, N. (1977, March). A photometric and polarmetric study of the Moon's surface II: On the possibility of the brightness fluctuations of the Moon. *Earth, Moon, and Planets*, 16, 199-213.

[251] Sekiguchi, N. (1979, June 18). Photometric and polarmetric observations of the Moon's surface (I). *Tokyo Astronomical Bulletin* (2nd Series, No. 257, pp. 2945-2951). Mitaka, Tokyo, Japan: Tokyo Astronomical Observatory.

[252] Sekiguchi, N. (1986, March 10). Photometric and polarmetric observations of the Moon's surface during 1984 and 1985. *Tokyo Astronomical Bulletin* (2nd Series, No. 275, pp. 3163-3184). Mitaka, Tokyo, Japan: Tokyo Astronomical Observatory.

[253] Cameron, (1978).

[254] Hilbrecht, H. & Küveler, G. (1984, February). Observations of lunar transient phenomena (LTP) in 1972 and 1973. *Earth, Moon, and Planets*, 30, 53-61. Provided by the NASA Astrophysics Data System.

[255] Sisterson, J. M., Kim, K., Caffee, M. W., & Reedy, R. C. (1997, March). Be-10 and Al-26 production in lunar rock 68815 – Revised production rates using new cross section measurements (pp. 326-327). Conference paper, *28th Annual Lunar and Planetary Science Conference*. Provided by the NASA Astrophysics Data System.

[256] Garrison, D. H., Rao, M. N., & Bogard, D. D. (1993). The SCR Ne-21 and Ar-38 in lunar rock 68815: The solar proton energy spectrum over the past 2 MYR (Part 2:

G-M, p. 521-522). Twenty-Fourth Lunar and Planetary Science Conference (SEE N94-16173 03-91). Provided by the NASA Astrophysics Data System. Garrison, D. H., Rao, M. N., Bogard, D. D., & Reedy, R. C. (1993, July). Determination of solar proton spectrum in lunar rock 68815. *Meteoritics*, **28**(3), 351. Provided by the NASA Astrophysics Data System. Sisterson, J. M., Kim, K., Caffee, M. W., & Reedy, R. C. (1997, March). Be-10 and Al-26 production in lunar rock 68815 – Revised production rates using new cross section measurements (pp. 326-327). Conference paper, *28th Annual Lunar and Planetary Science Conference*. Provided by the NASA Astrophysics Data System.

[257] Reedy, R. C. (1999, March). Variations in solar-proton fluxes over the last million years [Abstract no. 1643]. *30rd Annual Lunar and Planetary Science Conference*, Houston, TX. Provided by the NASA Astrophysics Data System. Reedy, R. C. (2002, March). Recent solar energetic particles: Updates and trends [Abstract no. 1938]. *33rd Annual Lunar and Planetary Science Conference*, Houston, TX. Provided by the NASA Astrophysics Data System.

[258] Ng, C. K., & Reames, D. V. (1994). Focused interplanetary transport of `1 MeV solar energetic protons through self-generated Alfven waves. *Astrophysics Journal*, **424**, 1032.

[259] Reames, D. V. (2001). SEP's: space weather hazard in interplanetary space (p. 101). In P. Song, H. J. Singer, & G. Siscoe (Eds.), *Space Weather* [Geophysical Monograph 125]. Washington, DC: American Geophysical Union.

[260] Kahler, S. W. (2003). Solar fast wind regions as sources of gradual 20MeV solar energetic particle events (AFRL-VS-HA,XC). (DTIS No. ADA423046). Brosius, J. (2003, November, 5). Characterization of large-scale solar corona. (DTIS No. ADA418377).

[261] Dollfus, A. & Bowell, E. (1971). Polarmetric properties of the lunar surface and its limitation. *Astronomy and Astrophysics*, **10**, 29-53. Provided by the NASA Astrophysics Data System.

[262] White, E. G. (1988). *America in Prophecy* [pp. 316-317].. Jemison, AL: Inspiration Books East, Inc. (Original work published 188 under the title *The Great Controversy*).

[263] Burnham, Robert (2000). *Great Comets* [p. 12]. New York: Cambridge University Press.

[264] Reynolds, M. D. (2001). *Falling Stars: A Guide to Meteors and Meteorites* [p. 44]. Mechanicsburg, PA: Stackpole Books.

[265] Yeomans, Donald K. (1991). *Comets: A Chronological History of Observation, Science, Myth, and Folklore* [p. 190]. New York: John Wiley & Sons.

[266] McKinley, D. W. R. (1961). Meteor. In *Encyclopedia Britannica* (Vol. 15, pp. 334-336). Chicago: Encyclopedia Britannica.

[267] Jewitt, D. (2000). Astronomy: Eyes wide shut. *Nature*, **403**, 145-148. Jewitt, D. (2000, May/June). Astronomy: Eyes wide shut. *Planetary Report*, **20**(3), 4.

[268] Reynolds, p. 45.

[269] Federer, C. A., (Ed.) (1962, February). *Sky & Telescope*, **23**(2), 64.

[270] Considine, D. M. (Ed.) (1976). *Van Nostrand's ScientificEencyclopedia* (5th ed.) [p. 1457]. New York: D. Van Nostrand Reinhold Company.

[271] Taylor, C. E. (1979). *World War III and the destiny of America* [p. 325]. Nashville, TN: Thomas Nelson Inc., Publishers.
[272] Lindsey, H. (1977). *There's a New World Coming* (5th printing) [pp. 93-97]. New York: Bantam Books, Inc.
[273] Lindsey, H. (1997). *Apocalypse Code* [pp. 108-112]. Palos Verdes, CA: Western Front, Ltd.
[274] Asada, T., Baba, H., Kawazoe, M., & Sugiura, M. (2001, January). An attempt to delineate very low frequency electromagnetic signals associated with earthquakes [abstract]. *Earth, Planets, and Space,* **53**(1), 55-62. (Originates from TERRAPUB, Japan)
[275] Whiston, W. (Trans.) (1960). *Josephus complete works* (Antiq., book VI, chap. II, sec. 2, p. 125). Grand Rapids, MI: Kregel Publications.
[276] Altschuler, M. D., (1969). Chapter 7. Atmospheric electricity and plasma interpretation of UFOs. In E. U. Condon (Sci. Dir.) & D. S. Gillmor (Ed.). *Scientific study of Unidentified Flying Objects* [pp. 740-741]. New York: A Bantam Books, Inc.
[277] Brookesmith, P. (1995). *UFO: The complete sighting* [pp. 17, 120, 126, 139, 157]. New York: Barnes and Nobel, Inc.
[278] Sturrock, P. A. (1999). *The UFO enigma: A new review of the physical evidence* [pp. 143-144]. New York: Warner Books, Inc.
[279] Teordorani, M. (2004). A long-term scientific survey of the Hassdalen phenomenon. *Journal of Sccientific Exploration,* **18**(2), 217-251.
[280] Sehra, A. K. & Whitlow, W. (2004). Propulsion and power for 21st century aviation. *Progress in Aerospace Sciences,* **40**, 199-235.
[281] Nelson, R. A. (2005, March last update). Airplanes. In Rex Research at http://www.rexresearch.com.
[282] Flying Disc, The (1954, December). *Air Intelligence Digest,* 7(12), 6-12. Retrieved April 18, 2003, from http://www.cufon.org/cufon/flydisc.htm.
[283] Hilton, W. F. (1958, April). Flying saucers – are they best for space flight? *Aircraft and Missiles Manufacturing* (pp. 50-51, 82).
[284] General Dynamics Corporation (1961, July). Pye Wacket. Feasibility test vehicle study. Summary. Volume I (Tech. Rep. No. TR-61-34-V1, ASD). (DTIS No. AD325216) Pictured in the Pye Wacket web site, retrieved August 25, 2003, from http://www.laesieworks.com/ifo/lib/PyeWacket.html.
[285] Blanchard, U. J. (1961, September). Landing characteristics of a lenticular-shaped re-entry vehicle (Tech. Rep. No. TN D-940). (DTIS No. AD263072).
[286] Oberto, R. J. (1962, October). Environmental control systems for manned space vehicles. Volume II: Appendix I, Missions, vehicles, and equipment (Tech. Rep. No. TR-61-240-V2-PT2, ASD). (DTIS No. AD333666).
[287] Wilson, J. (2000, November). America's nuclear flying saucer. *Popular Mechanics,* pp. 66-69, 71.
[288] Bennett, G. L. (1997, July). Some observations on avoiding pitfalls in developing future flight systems (AIAA 97-3209). Paper presented at the 33rd *AIAA/ASME/SAE/ASEE Joint Propulsion Conference & Exhibit,* Seattle, WA.

[289] Stevens, H. (2003). *Hitler's flying saucers: A guide to german flying discs of the second world war* [p. 127]. Kempton, IL: Adventures Unlimited Press.
[290] Coats, C. (2001). *Living energies.* (First print 1996). Dublin, Ireland: Gateway an imprint of Gill & Macmillan Ltd.
[291] Shauberger, V., & Coats, C (Trans. & Ed.) (2000). *The Energy Evolution – Harnessing Free Energy from Nature.* Dublin, Ireland: Gateway an imprint of Gill & Macmillon Ltd.
[292] Cook, N. (2002). *The hunt for zero point: Inside the classified world of antigravity technology* [pp. 204-222]. New York: Broadway Books.
[293] Coats, 287, 289, 292.
[294] Stevens, 121-128.
[295] Schauberger & Coats, 192, 195.
[296] Stearns, E. F. (1911, December). The Tesla turbine. *Popular Mechanics.* Retrieved on May 27, 2004, from the Lindsey's Technical Archive: http://www.lindsaybks.com/arch/turbine.
[297] Germano, F., Dorantes, M., Johnson, T. & Letourneau, P. E. (2002/2003). Prior historical works of International Turbine and Power, LLC (1999-2003). Retrieved June 17, 2004 from http://www.frankgermano.com/theturbine.htm.
[298] Miller, G. E., Etter, B. D. & Dorsi, J. M.)1990, February). A multiple disk centrifugal pump as a blood flow device [Abstract]. *IEEE Trans Biomedical Engineering,* **37**(2), 157-163. Miller, G. E. & Fink, R. (1999, June). Analysis of optimal design configurations for a multiple disk centrifugal blood pump [Abstract]. *Artificial Organs,* **23**(6), 559-565. Miller, G. E., Madigan, M. & Fink, R. (1995, July). A preliminary flow visualization study in a multiple disk centrifugal artificial ventricle [Abstract]. *Artificial Organs,* **19**(7), 680-684. Vermette, P., Thibault, J. & Laroche, G. (1998, September). A continuous and pulsatile flow circulation system for evaluation of cardiovascular devices [Abstract]. *Artificial Organs,* **22**(9), 746-752.
[299] See, for example: Electromagnetic pulse (2006). Retrieved June 14, 2006 from Wikipedia – the free encyclopedia, at http://en.wikipedia.org/eiki/Electromagnetic_pulse. Steinbruner, J. (1984, January). Launch under attack. *Scientific American,* **250**(1), 37-47.
[300] Rabinowitz, M. (1987). Effect of the fast nuclear electromagnetic pulse on the electric power grid nationwide: A different view. *IEEE Transmission Power Delivery* **PWRD-2,** 1199-1222. Preprint arXiv: physics/0307127.
[301] Ibid.
[302] Vittitoe, C. N., & Rabinowitz, M. (1988). Radiative reactions and coherence modeling in the high altitude electromagnetic pulse. *Physical Review A,* **37,** 1969-1977. Preprint arXiv: physics/0306028.
[303] Rabinowitz, M., Meliopoulos, A. P. S., Glytsis, E. N., & Cokkinides, G. J. (1992). Nuclear magnetohydrodynamic EMP, solar storms, and substorms. *International Journal of Modern Physics B,* **6**(20), 3353-3380. Preprint arXiv: physics/0307067.
[304] Plait, P. (2006, June). Death from the skies. *Sky & Telescope,* **111**(6), 30-34.
[305] LaViolette, P. (1983). *Galactic explosions, cosmic dust invasion, and climatic change*

(Ph.D. dissertation). Portland State University, Oregon. Provided by the NASA Astrophysics Data System.
306. LaViolette, P. (1987). Cosmic ray volleys from the galactic center and their recent impact on the Earth environment. *Earth, Moon, and Planets*, **37**, 241-286. Provided by the NASA Astrophysics Data System.
307. Phillips, T. (2003, September 12). Solar flares on steroids. Retrieved June 22, 2006 from http://science.nasa.gov/headlines/y2003/12sep_magnetars.htm.
308. Duncan, R. D. (2003, March). Magnetar's, soft gamma repeaters & very strong magnetic fields. Retrieved June 22, 2006 from http://solomon.as.utexas.edu/~duncan/magnetar.html.
309. Coburn, W., & Boggs, S. E. (2003, May 20). Polarization of the prompt =-ray emission from the =-ray burst of 6 December 2002. Preprint arXiv: astro-ph/0305377 v1.
310. Sanders, R. (2003, May 28). RHESSI uncovers secret to cataclysmic explosions known as gamma-ray bursts. Retrieved June 16, 2006 from http://www.berkeley.edu/news/media/releases/2003/05/28_gamma.shtml.
311. Rutledge, R. E., & Fox, D. B. (2004, January 26). Re-analysis of polarization in the =-ray flux of GRB 021206. Preprint arXiv: atro-ph/0310385 v2.
312. Boggs, S. E., Coburn, W., & Kalimci, E. (2006, February). Gamma-ray polarimetry of two X-class solar flares. *The Astrophysical Journal*, **638**(2), 1129-1139. Preprint arXiv: astro-ph/0510588.
313. Hurford, G. J., Krucker, S., Lin, R. P., Schwartz, R. A., Share, G. H., & Smith, D. M. (2006, June). Gamma-ray imaging of the 2003 October/November solar flares. *Astrophysical Journal*, **644**(1), L93-L96.
314. Yenne, B. (1985). *The encyclopedia of US spacecraft*. New York: Exeter Books, pp. 151-152, 170-172.
315. Palle, E., Butler, C. J., & O'Brien, K. O. (2004). The possible connection between ionization in the atmosphere by cosmic rays and low level clouds. *Journal of Atmospheric and Solar-Terrestrial Physics*, **66**, 1779-1790.
316. Perry, C. A. (2007). Evidence for a physical linkage between galactic cosmic rays and regional climate time series. *Advances in Space Research*, **40**, 353-364.
317. Cliver, E. W. (2006). The 1859 space weather event: Then and now. *Advances in Space Research*, **38**(2), 119-129.
318. Green, J. L.; Boardsen, S., Odenwald, S., Humble, J., & Pazamickas, A. (2006). Eye witness reports of the great auroral storm of 1859. *Advances in Space Research*, **38**(2), 145-154.
319. Whitehouse, D. (2005). *The Sun – A Biography*. West Sussex, England: John Wiley & Sons, Ltd.
320. Clarke, S. (2007). *The Sun Kings – The Unexpected Tragedy of Richard Carrington and the Take of How Modern Astronomy Began*. Princeton, NJ: Princeton University Press.
321. Cliver, pp. 119-129.
322. Clark, S. (2007). Astronomical fire: Richard Carrington and the solar flare of 1859. *Endeavour*, **31**(3), 104-109.
323. Green, pp. 145-154.

[324] Brooke, C. (1847). On the automatic registration of magnetometers, and other meteorological instruments, by photography. *Philosophical Transactions of the Royal Society of London – Part I* [pp. 69-77]. London, England: Richard and John E. Taylor.

[325] Cliver, E. W., & Svalgaard, L. (2004). The 1859 solar-terrestrial disturbance and the current limits of extreme space weather activity. *Solar Physics*, **224**, 407-422.

[326] Tyasto, M. I., Ptitsyna, N. G., Veselovsky, I. S., & Yakovchouk, O. S. (2009). Extremely strong geomagnetic storm of September 2-3, 1859, according to the archived data of observatories at the Russian network. *Geomagnetism and Aeronomy*, **49**(2), 153-162.

[327] Sabine, E. (1861). Major-General Sabine – Report of the Kew Committee. *Report of the Thirteenth Meeting of the British Association for the Advancement of Science; Held at Oxford in June and July 1860* [pp. xxxv-xxxvi]. London, England: John Murray.

[328] Chambers, G. F. (1889). *A Handbook of Descriptive and Practical Astronomy* [p. 14]. Oxford, England: The Clarendon Press.

[329] Hodgson, R. (1859, November). On a curious appearance seen in the Sun. *Monthly Notices of the Royal Astronomical Society*, **20**, 15-16. Provided by the NASA Astrophysics Data System.

[330] Carrington, R. C. (1859, November). Description of a singular appearance seen in the Sun on September 1, 1859. *Monthly Notices of the Royal Astronomical Society*, **20**, 13-15. Provided by the NASA Astrophysics Data System.

[331] Christopher D'Andrea (2006, December 15). *Analysis of Ground Level Muons, Solar Flares, and Forbush Decreases* (Dissertation) [p. 9]. Notre Dame, IN: Graduate School of the University of Notre Dame.

[332] Clark, *The Sun Kings*, pp. 110-111.

[333] Whitehouse, p. 130.

[334] San Francisco Earthquake History 1769-1879. Retrieved March 5, 2010 from http://www.sfmuseum.org/alm/quakes1.html.

[335] Guzzetti, F., et al (2009, September). Central Italy seismic sequences-induced landsling: 1997-1998 Umbria-Marche and 2008-2009 L'Aquila cases. In, C-T. Lee (ed.), *The Next Generation of Research on Earthquake-induced Landslides: An International Conference in Commemoration of 10*th *Anniversary of the Chi-Chi Earthquake*. Taoyuan County, Taiwan (R.O.C.): National Central University.

[336] Taber, S. (1922, February-March). The great fault troughs of the Antilles. *The Journal of Geology*, **30**(2), 89-114.

[337] Woodring, W. P. (1924). Earthquakes. In, W.P. Woodring, J. S. Brown, W. S. Burbank (contributors). Department of Public Works, *Geology of the Republic of Haiti* [pp. 338-349]. Baltimore, MD: Lord Baltimore Press.

[338] McCann, W. R. (2007, May). Estimating the threat of tsunamigenic earthquakes and earthquake induced-landslide tsunami in the Caribbean. *American Geophysical Union*, Spring Meeting 2007, Abstract #T24A-03, pp. 43-65.

[339] Golub, L., & Pasachoff, J. M. (2001). *Nearest Star: The Surprising Science of our Sun* [p. 232]. Cambridge, MA: Harvard University Press.

[340] Whitehouse, D. (2005). *The Sun – A Biography* [pp. 238-240]. West Sussex, England: John Wiley & Sons, Ltd.
[341] Zeller, E. J., Dreschhoff, G. A. M., & Laird, C. M. (1986). Nitrate flux on the Ross Ice Shelf, Antarctica and its relation to solar cosmic rays. *Geophysical Research Letters*, **13**(12), 1264-1267.
[342] Dreschhoff, G. A. M., & Zeller, E. J. (1990, June). Evidence of individual solar proton events in Antarctic snow. *Solar Physics*, **127**, 333-346.
[343] Shea, M. A., Smart, D. F., McCracken, K. G., Dreschhoff, G. A. M., & Spence, H. E. (2006). Solar proton events for 450 years: The Carrington event in perspective. *Advances in Space Research*, **38**(2), 323-238.
[344] McCracken, K. G., Dreschoff, G. A. M., Zeller, E.J., Smart, D. F., & Shea, M. A. (2001, October). Solar cosmic ray events for the period 1561-1994 I.
[345] Dreschhoff, G. A. (2002). Stratigraphic evidence in polar ice of variations in solar activity: Implications for climate? *2002 PACLIM Conference Proceedings* [pp. 23-33].
[346] Bullard, F. M. (1984). *Volcanoes: In History, in Theory, in Eruption* (2nd ed.) [p. 334]. Austin, TX: University of Texas Press.
[347] Wilson, R. H. (1960, August). Magnetic field effects on artificial satellites. *Sky and Telescope*, **20**(2), 77-79.
[348] Earth's Solid Core Outruns Its Crust (1996, October). *Sky & Telescope*, **92**(4), 10.
[349] Stephenson, pp. 513-517.
[350] Kokus, M. (2002). Alternate theories of gravity and geology in earthquake prediction. In M. R. Edwards (Ed.) *Pushing gravity: New perspectives on LeSage's theory of gravitation* (pp. 285-302). Montreal, Quebec, Canada: C. Roy Keys Inc.
[351] Murray, J. (1840). *Truth of Revelation* [pp. 216-217]. London, UK: William Smith. Cited by T. Mortenson (2002, February 5). Glenn Morton: Who's misrepresenting history? Retrieved February 10, 2009, from http://www.creationontheweb.org/content/view/4237.
[352] Brown, W. (2008). *In the beginning: Compelling evidence for creation and the flood* (8th ed) [pp. 117-121]. Phoenix, AZ: Center for Scientific Creation.
[353] Brown (2008), 228-235.
[354] Brown (2008), 263-293.
[355] Brown (2008), 295-313.
[356] Brown (2008), 122-125.
[357] Brown (2008), 122.
[358] Hunter, M. J. (2000, August). Scriptural constraints on the variation of water level during the Genesis flood. *Journal of Creation*, **14**(2), 91-94
[359] Brown (2008), 126-128.
[360] Brown (2008), 125.
[361] Mate, B. R., Gisiner, R., & Mobley, J. (1998). Local and migratory movements of Hawaiian humpback whales tracked by satellite telemetry. *Canadian Journal of Zoology*, **76**(5), 863-868. (DTIC No. ADA355039.
[362] Kohler, N. E., Turner, P. A., Hoey, J. J., Natanson, L. J., & Briggs, R. (2002). Tag and recapture data for three pelagic shark species: Blue shark (prionace glauca),

shortfin mako (isurus xyrinchus), and porbeagle (lamna nasus) in the North Atlantic Ocean. *ICAAT Collective Volume of Scientific Papers*, **54**(4), 1231-1260.

[363] Humphreys, D. R. (1984, December). The creation of planetary magnetic fields. *Creation Research Society Quarterly*, **21**(3). Retrieved September 26, 2002, from Creation Research Society web site.

[364] Hutton, V. R. S., Gought, D. I., Dawes, G. J. K., & Travassos, J. (1987, July). Magnetotelluric soundings in the Canadian Rocky Mountains. *Geophysical Journal International*, **90**(1), 245-263.

[365] Meade, C. & Jeanloz, R. (1991, April 5). Deep-focus earthquakes and recycling of water into the Earth's mantle [Abstract]. *Science*, **252**, 68-72.

[366] Wei, W., Unsworth, M., Jones, A., et la (2001, April 27). Detection of widespread fluids in the Tibetan Crust by magnetotelluric studies. *Science*, **292**(5517), 716-719. See also, Wei, W., Jin, S., Ye, G., Deng, M., Jing, J., Unsworth, M., & Jones, A. G. (2010, February). Conductivity structure and rheological property of lithosphere in Southern Tibet inferred from super-broadband magnetotelluric sounding. *Science China Earth Sciences*, **53**(2), 189-202.

[367] Li, S., Unsworth, M. J., Booker, J. R., Wei, W., Tan, H., & Jones, A. G. (2003, May). Partial melt or acqueous fluid in the mid-crust of southern Tibet? Constraints from INDEPTH magnetotelluric data. *Geophysical Journal International*, **153**(2), 289-304. See also, Unsworth, M. (2010). Magnetotelluric studies of active continent-continent collisions. *Surveys in Geophysics*, **31**, 137-161.

[368] Patro, B. P. K., Brasse, H., Sarma, S. V. S., & Harinarayana, T. (2005). Electrical structure of the crust below the Deccan Flood Basalts (India), inferred from magnetotelluric soundings. *Geophysical Journal International*, **163**, 931-943.

[369] Chen, C., Chen, C., Chiang, E., et al (2007, March). Crustal resistivity anomalies beneath central Taiwan imaged by broadband magnetotelluric transect. *Terrestrial, Atmospheric, and Oceanic Sciences*, **18**(1), 19-30.

[370] Helffrich, G. R., & Wood, B. J. (1996). 410 km discontinuity sharpness and the form of the olivine $\alpha^=$ phase diagram: Resolution of apparent seismic contradictions. *Geophysical Journal International*, **126**, F7-F12.

[371] van der Meijde, M., Marone, F., van der Lee, S, &. Giardini, D. (2002, December). Seismological evidence for the presence of water near the 410 km discontinuity. *American Geophysical Union, Fall Meeting 2002*, Abstract #S52C-03.

[372] van der Meijde, M., Marone, F., Giardini, D., & van der Lee, S. (2003, June 6). Seismic evidence for water deep in Earth's upper mantle. *Science*, **300**(5625), 1556-1558.

[373] Meade, C. & Jeanloz, R. (1991, April 5). Deep-focus earthquakes and recycling of water into the Earth's mantle [Abstract]. *Science*, **252**, 68-72.

[374] Green, H. W., & Green, H. W. (2001, December). Physical mechanisms for earthquakes at intermediate depths [Abstract #S42D-03]. *American Geophysical Union, Fall Meeting 2001*.

[375] Shaltout (1999), 502-506.

[376] Waltham, T. (1974). *Caves* [p. 7]. New York: Crown Publishers, Inc.

[377] Alpha, T. R., Galloway, J. P., & Tinsley, J. C., III (1997, October 10). Karst

topography – Computer animations and paper model. *Open file report 97-536-A* [p. 6]. Menlo Park, CA: U.S. Geological Survey.
[378] Alpha, p. 5.
[379] Kao, A. (1985, April). *Literature survey of underground construction methods for application to hardened facilities* (TR-M-85/11, CERL) [p. 7]. (DTIC No. ADA-155212).
[380] Waltham, p. 189.
[381] Hostettler, B., & Widmer, K. (1992, June 8). Shelters for the population as protective measures against chemical warfare agents. (DTIC No. ADD752631).
[382] Kearny, C. (1990, September). *Nuclear war survival skills* (Updated and expanded 1987 Edition) [p. 6]. Cave Junction, OR: Oregon Institute of Science and Medicine.
[383] Kao, p. 7.
[384] Kao, p. 9.
[385] Tarkoy, P. J., & Byram, J. E. (1991, January). The advantages of tunnel boring: A qualitative/quantitative comparison of D&B and TBM excavation. *Hong Kong Engineer*. Retrieved June 14, 2004 from http://home.comcast.net/~geoconsol/TBM-DB-1.pdf.
[386] Sauder, R. (1995). *Underground bases and tunnels: What is the government trying to hide?* Kempton, IL: Adventures Unlimited Press.
[387] Sauder, R. (2001). *Underwater and underground bases: Surprising facts the government does not want you to know!* Kempton, IL: Adventures Unlimited Press.
[388] Sauder, 2001, pp. 226-253.
[389] Tarkoy, P. J. (1995, October). Comparing TBMs with drill+blast excavation. *Tunnels & Tunnelling*. Retrieved March 25, 2005 from http://www.geoconsol.com/.
[390] Tarkoy, P. J., & Byram, J. E. (1991, January). The advantages of tunnel boring: A qualitative/quantitative comparison of D&B and TBM excavation. *Hong Kong Engineer*. Retrieved June 14, 2004 from http://home.comcast.net/~geoconsol/TBM-DB-1.pdf.
[391] Reich, R. B., & Dear, J. A. (1996). *Underground construction (tunneling)* (OSHA-3115). Washington, DC: U.S. Department of Labor, OSHA. (DTIC No. ADA364853).
[392] Tarkoy, P. J. (1988, September). The stuff that claims are made of. *World Tunnelling*, 1(3), 249-253.
[393] Oggeri, C., & Ova, G. (2004). Quality in tunnelling: ITA-AITES working group 16 final report. *Tunnelling and Underground Space Technology*, **19**, 239-272.
[394] Tarkoy, P. J., & Wagner, J. (1988, October). Backing up a TBM. *Tunnels & Tunnelling*. (pp. 27-32) Retrieved March 25, 2005 from http://www.geoconsol.com/.
[395] Morris, D. A., Crawford, K. C., Kloesel, K. A., & Wolfinbarger (2001, September 9). OK-FIRST: A meteorological information system for public safety. *Bulletin for the American Meteorological Society*, **82**(9), 1911-1923.
[396] Dorman, L. I. (2008). Natural hazards for the Earth's civilization from space,

1. Cosmic ray influence on atmospheric processes. *Advances in Geosciences*, **14**, 281-286.
[397] Kao, p. 37.
[398] Sterling, R. L., & Godard, J-P. (2003-2004). Geoengineering considerations in the optimum use of underground space (ITA – Technical Reports) [pp. 4-6]. Lausanne, Switzerland: ITA-AITES. Retrieved June 14, 2004 from the ITA-AITES web site http://www.ita-aites.org.
[399] v. Braun, W. & Ordway, F. I., III (1976). *The rockets' red glare* [p. 160]. Garden City, New York: Anchor Press/Doubleday.
[400] Sepp, Eric (2000, May). Deeply buried facilities: Implications for military operations. *Occasional Paper No. 14* [p. 1]. Maxwell Air Force Base, AL: Air University, Air War College, Center for Strategy and Technology.
[401] Cornwall, J., Despain, A., Eardley, D., Garwin, R., & Hammer, D. (1999, April 12). *Characterization of underground facilities* (Tech. Rep. No. JSR-97-155). (DTIC No. ADA-363359).
[402] Secretary of Defense and Secretary of Energy (2001, July). *Report to Congress on the defeat of hard and deeply buried targets* [p. 8]. Retrieved May 6, 2004 from http://www.nukewatch.org/facts/nwd/HiRes_Report_to_Congress_on_the_Defeat.pdf.
[403] Moore, T. (2003, June 6). *Does the United States need to develop a new nuclear earth penetrating weapon?* (DTIC No. ADA416968).
[404] *Pictorial History of World War II: Volume I: The War in Europe* (1955). United States: Veterans' Historical Book Service, Inc., p. 364.
[405] Ley, W. (1969). *Rockets, missiles, and men in space* [p. 277]. New York: The New American Library, Inc.
[406] Cornwall, J., Despain, A., Eardley, D., Garwin, R., & Hammer, D. (1999, April 12). *Characterization of underground facilities* (Tech. Rep. No. JSR-97-155) [p. 47]. (DTIC No. ADA-363359).
[407] Rumsfeld: In the end, Saddam 'not terribly brave.' (2003, December 14). Retrieved March 25, 2005 from CNN.com
[408] Mackie, R. L. (1999, February). *Imaging of underground structure using HAARP* (AFRL-VS-TR-1999-1511). (DTIC No. ADA398268). Ganguly, S. (2000, November 20). *Experimental demonstration of underground structure characterization using sensitive magnetic sensors* (AFRL-VS-TR-2001-1606). (DTIC No. ADA406535).
[409] Butler, D. K. (2000, December). *Assessment of microgravimetry for UXO detection and discrimination* (ERDC/GSL TR-00-5). (NTIS No. ADA-387603). Maier, M. W. (2002, January 15). *Underground structures and gravity gradiometry* (TR-2001 (3000)-1). (DTIC No. ADA400252). Baker, R. M. L., Jr. (2005, February 6). Applications of high-frequency gravitational waves (HFGWs). *AIP Conference Proceedings*, **746**(1), 1306-1314.
[410] Yurkewicz, K. (2007, December). Across the ocean, yet close to home. *Symmetry*, **4**(10), 10-13.
[411] Harris, D. (2010, February). Science opportunities deep underground. *Symmetry*,

7(1), 2. Harris' editorial introduces the special issue that has been dedicated to underground scientific laboratories.

[412] Roberts, A., & Goldstein, M. (2000). Under the weather: Space weather. The magnetic field of the heliosphere [p. 67]. Greenbelt, MD: NASA Goddard Space Flight Center. (NTIS No. N20020090260).

[413] Stern, D. P., & Peredo, M. (2001, November 25). Solar energetic particles. Retrieved June 25, 2004 from the Exploration of the Earth's Magnetosphere web site

[414] Karpov, S. N., Miroshnichenko, L. I., & Vashenyuk, E. V. (1998, Settembro-Ottobre). Muon bursts at the Bakson Underground Scintillation Telescope during energetic solar phenomena. *Il Nuovo Cimento*, **21**-C(5), 551-573.

[415] Mavromichalaki, H., Papaioannou, A., Petrides, A., Assimakopoulos, B., Sarlanis, C., & Souvatzoglou, G. (2005). Cosmic ray event related to solar activity recorded at the Athens neutron monitor station for the period 2000-2003. *International Journal of Modern Physics A*, **20**(29), 6714-6716.

[416] Dvornikov, V. M., Lukovnikova, A. A., Sdobnov, V. E., Petukhov, S. I., & Starodubstev, S. A. (2008). Variation of angular distribution of cosmic rays during GLE on January 20, 2005. *Proceedings of the 30th International Cosmic Ray Conference*, **1**(SH), 197-200.

[417] Storini, M., Cordaro, E. G., & Parisi, M. (2007). The December 2006 GLE event as seen from LARC, SVIRCO and OLC. *30th International Cosmic Ray Conference* [ICRC 2007 Proceedings – Pre-Conference Edition]. Mérida, Mexico.

[418] Reames, D. V. (1999). Solar energetic particles: is there time to hide. *Radiation Measurements*, **30**, 297-308. Reames, D. V. (2001). SEP's: space weather hazard in interplanetary space (p. 101). In P. Song, H. J. Singer, & G. Siscoe (Eds.), *Space Weather* [Geophysical Monograph 125]. Washington, DC: American Geophysical Union. Reames, D. V. (2002). Solar energetic particles. *Space Radiation* (Japan), **3**, 69.

[419] Wilson, J. W., Goldhagen, P., Friedberg, W., De Angelis, G., Clem, J. M., Copeland, K., & Bidasaria, H. B. (2005, November). *Atmospheric Ionizing Radiation and Human Exposure*. [NASA/TP-2005-213935]. Hampton, VA: NASA Langley Research Center. See Table 14. Annual neutron collective dose equivalents for various exposed groups.

[420] Wilson, J. W., Cucinotta, F. A., Kim, M. Y., Shinn, J. L., Jones, T. D., & Chang, C. K. (1998, July 12-19). Biological response to SPE exposures. *32nd COSPAR Scientific Assembly*, Nagoya, Japan. Retrieved March 22, 2005 from the Langley Technical Reports Server at http://techreports.larc.nasa.gov/1trs/cit.html.

[421] Friedberg, W., & Copeland, K. (2005, February 14). What aircrews should know about their occupational exposure to ionizing radiation. Oklahoma City, OK: Federal Aviation Administration, Civil Aerospace Medical Institute. Retrieved March 14, 2005, from http://www.cami.jccbi.gov/AAM-600/Radiation/trainin. Rafnsson, V., Hrafnkelsson, J., Tulinius, H., Sigurgeirsson, B., & Olafsson, J. H. (2003). Risk factors for cutaneous malignant melanoma among aircrews and a random sample of the population. *Occupational & Environmental Medicine*, **60**(11), 815-820. Whelan, E. A. (2003). Cancer incidence in airline cabin crew.

Occupational & Environmental Medicine, **60**(11), 805-806. Townsend, L. W. (2001, March). Radiation exposure of aircrews in high altitude flight. *Journal of Radiological Protection*, **21**(1), 5-8. Nicholas, J. S., Copeland, K., Duke, F. E., Friedberg, W., & O'Brien, K., III (2000, October). Galactic cosmic radiation exposure of pregnant aircrew members II – Final report (Tech. Rep. No. DOT/FAA/AM-00/33). (NTIS No. PB2001-102921).

[422] Rafnsson, V., Sulem, P., Tulinius, & Hrafnkelsson, J. (2003). Breast cancer risk in airline cabin attendants: A nested case-control study in Iceland. *Occupational & Environmental Medicine*, **60**(11), 807-809. Haldorsen, T., Reitan, J. B., & Tveten, U. (2003). Cancer incidence among Norwegian airline cabin attendants. *International Journal of Epidemiology*, **30**(4), 825-830.

[423] Pukkala, E., Aspholm, R., Auvinen, A., Eliasch, H., Gundestrup, M., Haldorsen, T., et al (2002, September 14). Incidence of cancer among Nordic airline pilots over five decades: Occupational cohort study. *BMJ*, **325**(736), 567.

[424] Linnersjö, A., Hammer, N., Dammström, B.-G., Johansson, M., & Eliasch, H. (2003). Cancer incidence in airline cabin crew: Experience from Sweden. *Occupational & Environmental Medicine*, **60**(11), 810-814.

[425] Poppe, B. (2005, March 1). NOAA space weather scales. Retrieved April 7, 2005 from the Space Environment Center web site at http://www.sel.noaa.gov/NOAAscales/#SolarRadiationStorms.

[426] Acute Radiation Syndrom (A fact sheet for physicians) (2002, April). National Center for Environmental Health/Radiation Studies Branch (Draft).

[427] Dimitrova, S. (2007). Geomagnetic indices variations and human physiology. *Sun and Geosphere*, **2**(2), 84-87.

[428] Stoupel, E. (2006). Cardiac arrhythmia and geomagnetic activity. *Indian Pacing and Electrophysiology Journal*, **6**(1), 49-53.

[429] Babayev, E. S. (2008). Solar and geomagnetic activities and related effects on the human physiological and cardio-health state: Some results of Azerbijani and collective studies. In, A. Hady & M. I. Wanas (eds.), *1st Middle East and Africa IAU-Regional Meeting Proceedings MEARIM*, **1**, 171-177.

[430] Stoupel, E., Babayev, E. S., Shustarev, N., Abramson, E., Israelevich, P., & Sulkes, J. (2009, November). Traffic accidents and environmental physical activity. *International Journal of Biometeorology*, **53**(6), 523-534.

[431] Dorman, L. I. (2005). Space weather and dangerous phenomena on the Earth: Principles of great geomagnetic storm forcasting by online cosmic ray data. *Annales Geophysicae*, **23**, 2997-3002.

[432] Green, J. L.; Boardsen, S., Odenwald, S., Humble, J., & Pazamickas, A. (2006). Eye witness reports of the great auroral storm of 1859. *Advances in Space Research*, **38**(2), 145-154. See pages 27-28.

[433] Shiga, D. (2007, August 21). Vibrations on the Sun may 'shake' the Earth. *New Scientist*. Retrieved June 1, 2010, from http://www.newscientist.com/article/dn12520-vibrations-on-the-sun-may-shake-the-earth.html.

[434] Mavromichalaki, H., Papaioannou, A., Petrides, A., Assimakopoulos, B., Sarlanis, C., & Souvatzoglou, G. (2005). Cosmic ray event related to solar activity recorded

at the Athens neutron monitor station for the period 2000-2003. *International Journal of Modern Physics A*, **20**(29), 6714-6716.

[435] Stoupel, E. (1999). Effect of geomagnetic activity on cardiovascular parameters. *Journal of Clinical and Basic Cardiology*, **2**(1), 34-40.

[436] Palmer, S. J., Rycroft, M. J., & Cermack, M. (2006). Solar and geomagnetic activity, extremely low frequency magnetic and electric fields and human health at the Earth's surface. *Survey of Geophysics*, **27**, 557-595.

[437] Mavromichalaki, H., Papailiou, M., Dimitrova, S., Babayev, E. S., & Mustafa, F. R. (2008, September). Geomagnetic disturbances and cosmic ray variations in relation to human cardio-health state: A wide collaboration. In, P. Király, K. Kudela, M. Stehlík, and A. W. Wolfendale (eds.), *Proceedings of the 21st European Cosmic Ray Symposium* [pp. 351-356]. Košice, Slovakia: Slovak Academy of Sciences, Institute of Experimental Physics.

[438] Štetiarová, J., Dzvoník, O., Kudela, K., & Daxner, P. (2008, September). Biochemical parameters of human health monitored in season of low and high solar activity. In, P. Király, K. Kudela, M. Stehlík, and A. W. Wolfendale (eds.), *Proceedings of the 21st European Cosmic Ray Symposium* [pp. 153-157]. Košice, Slovakia: Slovak Academy of Sciences, Institute of Experimental Physics.

[439] Stoupel, E. Birk, E., Kogan, A,. Klinger, G., Abramson, E., Israelevich, P., Sulkes, J., & Linder, N. (2009). Congenital heart disease: Correlation with fluctuations in cosmophysical activity, 1995-2005. *International Journal of Cardiology*, **135**, 207-210.

[440] Johansson, O. (2009). Disturbance of the immune system by electromagnetic fields – A potentially underlying cause for cellular damage and tissue repair reduction which could lead to disease and impairment. *Pathophysiology*, **16**(2), 157-177.

[441] Stoupel, E., Domarkienė, S., Radišauskas, R., Bernotienė, G., Abramson, E., Israelevich, P., & Sulkes, J. (2005). Variants of acute myocardial infarction in relation to cosmophysical parameters. *Seminars in Cardiology*, **11**(2), 51-55.

[442] Stoupel, E., Tamoshiunas, A., Radishauskas, R., Bernotiene, G., Abramson, E., & Sulkes, J. (2010). Acute myocardial infarction (AMI) and intdermediate coronary syndrome (ACS). *Health*, **2**(2), 131-136.

[443] Heart Rhythm Society (2004). *Sudden Cardiac Death*. Retrieved April 29, 2010, from www.HRSpatients.org.

[444] Babayev, E. S., Allahverdiyeva, A. A., Mustafa, F. R., & Shustarev, P. N. (2007). An influence of changes of heliophysical conditions on biological systems: Some results of studies conducted in the Azerbaijan National Academic of Sciences. *Sun and Geosphere*, **2**(1), 48-52.

[445] Stoupel, E., Kalėdienė, R., Petrauskienė, J., Starkuvienė, S., Abramson, E., Israelevich, P., & Sukes, J. (2007). Monthly deaths number and concomitant environmental physical activity: 192 months observation (1990-2005). *Sun and Geosphere*, **2**(2), 78-83.

[446] Gigolashvili, M., Ramishvili, G., Janashia, K., Pitiurishvili, P., Gigolashvili, M., & Kukhianidze, V. (no date). Possible Dependence of the Sign Changing

of the Polarity of Interplanetary Magnetic Field on Complications of Various Arrhythmias. Retrieved February 19, 2010 from www.boun.edu.tr.

[447] Stoupel, E., Kusniec, J., Mazur, A., Abramson, E., Israelevich, P., & Strasberg, B. (2008). Timing of life-threatening arrhythmias detected by implantable cardioverter-defibrillators in relation to changes in cosmophysical factors. *Cardiology Journal*, **15**(5), 437-440.

[448] Papailiou, M., Mavromichalaki, H., Vassilaki, A., Kelesidis, K. M., Mertzanos, G. A., & Petropoulos, B. (2009, February 16). Cosmic ray variations of solar origin in relation to human physiological state during the December 2006 solar extreme events. *Advances in Space Research*, **43**(4), 523-529.

[449] Mavromichalaki (2008, September), pp. 351-356. Mavromichalaki, H., Papaioannou, A., Petrides, A., Assimakopoulos, B., Sarlanis, C., & Souvatzoglou, G. (2005). Cosmic ray event related to solar activity recorded at the Athens neutron monitor station for the period 2000-2003. *International Journal of Modern Physics A*, **20**(29), 6714-6716.

[450] Stoupel, E., Kalėdienė, R., et al (2007), pp. 78-83.

[451] Fogel, D. H., & Righthand, N. (1964). The effect of meteorological phenomena on acute myocardial infarction in Stamford, Connecticut. *British Heart Journal*, **26**, 255-260.

[452] Stoupel, E. (2006). Cardiac arrhythmia and geomagnetic activity. *Indian Pacing and Electrophysiology Journal*, **6**(1), 49-53.

[453] Stoupel, E., Tamoshiunas, A., Radishauskas, R., Bernotiene, G., Abramson, E., & Sulkes, J. (2010). Acute myocardial infarction (AMI) and intdermediate coronary syndrome (ACS). *Health*, **2**(2), 131-136.

[454] Stoupel, E., Babayev, E. S., Mustafa, F. R., Abramson, E., Israelevich, P., & Sulkes, J. (2006). Clinical cosmobiology – Sudden cardiac death and Daily/Monthly geomagnetic, cosmic ray and solar activity – the Baku study (2003-2005). *Sun and Geosphere*, **1**(2), 13-16.

[455] Marusek, J. A. (2007). Solar storm threat analysis [p. 20]. Bloomfield, IN: Impact. Retrieved February 11, 2010 from http://citeseerx.ist.psu.edu/viewdoc/downloa d?doi=10.1.1.129.6527&rep=rep1%type=pdf.

Made in the USA
Middletown, DE
01 March 2017